Schriften
der
Deutschen Shakespeare-Gesellschaft
Band I

Shakespeares Belesenheit

Eine Abhandlung
über Shakespeares Lecture und die unmittelbaren
Quellen seiner Werke

von

H. R. D. Anders
B.A. (Univ. of the Cape of Good Hope), Ph.D. (Berlin Univ.)

Berlin
Druck und Verlag von Georg Reimer
1904.

SHAKESPEARE'S BOOKS

A DISSERTATION ON SHAKESPEARE'S READING AND THE IMMEDIATE SOURCES OF HIS WORKS

BY

H. R. D. ANDERS

B.A. (UNIV. OF THE CAPE OF GOOD HOPE), PH.D. (BERLIN UNIV.)

BERLIN

PUBLISHER & PRINTER GEORG REIMER

1904.

TO

THE UNIVERSITY OF THE CAPE OF GOOD HOPE

my *alma mater*.

SHAKESPEARE'S BOOKS.

CONTENTS.

CHAPTER 1.

SHAKESPEARE AND THE CLASSICS.

CHAPTER 2.

MODERN CONTINENTAL LITERATURE.

Chapter 3.

The English Non-Dramatic Polite Literature.

Chapter 4.

The English Drama.

Chapter 5.

Popular Literature.

Chapter 6.

The Bible and the Prayer Book.

Chapter 7.

Shakespeare's Earth and Heaven.

Additional Notes.

PREFACE.

It is my pleasant duty to thank my friends, my helpers, and all those who have furthered this work. My heartfelt thanks are due in the first place to Mrs. Marianne v. Koenen, a truly fine-cultured and *gentle* woman, to whom I am most deeply indebted. The theme was suggested to me by Professor Doctor Brandl, a large-hearted and large-minded man, whose original lectures on the English Drama and on Shakespeare I had heard. He induced the German Shakespeare Society, which he serves with devoted enthusiasm, to give its support towards a proper completion of the work. The readiness with which this noble society responded has laid me under lasting obligations.

I have to thank right cordially Mr. P. A. Daniel for his most generous consent to read through the proofs with me and for his friendly criticism. I am further indebted to the Rev. J. W. Ebsworth, our first living authority on English ballads, who, despite weakness and illness against which he had to battle, looked through the fifth chapter with care and favoured me (and the reader) with many notes from the abundant store of his knowledge of old ballads and songs. My best thanks are also due to Miss Lucy Toulmin Smith, of Oxford; to Mr. E. S. Dodgson who suggested some alterations; to Dr. Furnivall; to Dr. Garnett; to Herr Kawerau, and others. I have also to thank the trustees of the Libraries for the great liberality and hospitality extended to me,—the Royal Library of Berlin, the British Museum Library (a paradise for scholars), and the Bodleian Library at Oxford. I was also allowed a liberal use of the choice collection of books in the English Seminary Library at Berlin, which owes its existence to the efforts of Professor Brandl, and of which he is justly proud.

The present work was intended to serve as an introduction to a new edition of Collier-Hazlitt's "Shakespeare's Library". This explains the tendency, plan, and scope of the dissertation. It explains, for example, why Cinthio, on p. 66, occupies only four lines, while Rabelais, whose influence on Shakespeare is inconsiderable, takes up two full pages. In an edition of "Shakespeare's Library" the importance of the former author in relation to 'Othello' would be brought home by reading Cinthio's novel; whereas of Rabelais's connexion with Shakespeare, which has been treated exhaustively on pp. 56—7, nothing further need be said. The amount of space occupied is therefore not always a true indicator of the quality or importance, either in this or in any other book. But I have generally tried to see and present things in their true proportions.

Mr. Daniel pointed out to me the disproportionate amount of space I devoted to a discussion of the sources of Cymbeline (pp. 60—64). Perhaps I should have made it part of the appendix to the chapter. The average reader may omit this section containing controversial matter. A longer discussion was occasioned by Ohle's work, which has influenced some commentators.

In the fifth chapter I have given many bibliographical details. This is due to the fact that the materials are widely scattered. A collective volume containing all the old ballads and songs, quoted or referred to by Shakespeare, together with the old music, and a short analysis of the plots of the romances, etc., in question, will be published by the German Shakespeare Society. This work which I refer to as the "later volume" (cf. p. 155) will form part of "Shakespeare's Library" (cf. p. 5).

Mr. Daniel said to me, that I had given too many references throughout. I admit I have given many. But they are not superfluous. References, generally speaking, are of two kinds. The one sort serves to shuffle off the shoulder of the writer the responsibility of his assertion. "If you", he means, "want to bother about investigating this question, you can study the article there or there. Do it if *you* like. As for me, I shall not tell you whether I read it; but I know about it". The other sort of reference is given by the writer who has honestly read the work referred to himself, and who refers to it, if it is deserving of mention. He gives *in nuce* what he considers important for his purpose and only refers the reader to it for further information, or as the source whence he derived his knowledge. Nor is a writer of this sort addicted to the vice of giving exhaustive bibliographical information which only serves to puzzle the reader. But he prefers to drown worthless books deeper than plummet did ever sound. In fine, an ideal writer is not one who imposes his learning on the reader, but one who does the *work for him.*

The immense literature which centres around the name of Shakespeare renders a work of the present nature rather trying. It meant tough fighting to grapple with this sea of books which threatened to drown all *independence of thought.* For it has been my constant aim not to accept a statement without convincing myself of its truth. I have tried hard to lift the subject-matter beyond the sphere of mere learned cumbrousness. And I have endeavoured, as much as possible, to exclude dubious matter and to avoid all guesses. The reader may form an idea of the extent of the Shakespeare literature from the bare fact that a very able and hard working scholar, Mr. Furness, is spending his *whole life* in collecting all the more important comments on the great poet, without the prospect of finishing the work singlehanded. He is now about seventy years of age and has only finished 13 out of the 37 plays. But by the time it is finished it will no longer be up to date, because of the many excellent recent editions by Dowden, Gollancz, Brandl, Craig, Hart, Herford, etc., etc., not to speak of the ever-increasing Shakespeare-criticism published every year. An act of Parliament ought to be passed forbidding the publication of new comments, until Mr. Furness's New Variorum Edition is finished and all the material thoroughly digested.—But, happily, it is not necessary to know all the comments for an intelligent understanding of Shakespeare, any more than it is requisite to acquire all the theological learning for a healthy understanding of the Bible. It is sometimes better *not* to know it. Shakespeare speaks to the Shakespeare in us, and unless we have this apprehensive and sympathetic soul, no comment will bring him nearer to us.

A long stay on the Continent (such as mine was), however beneficial in other respects, can hardly be said to improve one's style. Any defects in this direction in the present work will, I trust, be condoned on other grounds.

The quotations in this work are nearly all given at first hand. It involved much labour, but it was worth it. The quotations in Malone's Variorum Edition which have been repeated without change by numerous commentators are surprisingly inaccurate. The old orthography has been generally adhered to; but the typographical irregularities respecting *u* and *v* have been disregarded.

I have taken into consideration all the works of 'Shakespeare' included in the Globe edition, from which I quote. We are, however, safe in asserting that Shakespeare did not write every line contained in this edition. Pericles (not in the first Folio) is not his work in its entirety. He is not the sole author of Henry VIII. and 1. Henry VI. Nor do Timon and Cymbeline seem to be wholly his. We have a few interpolations in Macbeth and perhaps elsewhere.[1] But the main portion of the works issued under Shakespeare's name in 1623 is his. Of course, this is a truism. But perhaps it is well to state it.

In some few cases we may feel some difficulty, whether this or that item should be treated as Shakespearean, as I have done in the present treatise, or whether it should not be regarded as having been taken over from an older substructure of the play in question, or added by another hand. But, on the whole, the older *extant* plays, known to, or rehandled, by Shakespeare, give us a clear insight into his manner of working and thinking; and, in general, one piece of evidence we have adduced so strongly supports the other, that the main results of our investigations must be regarded as perfectly safe.

It is exceedingly improbable that Shakespeare was the owner of a private library of large dimensions. In the absence of public libraries in those days, it becomes natural to ask where the poet found the volumes he required. It has been suggested that the libraries of his patron, the Earl of Southampton, of Jonson, Camden, and of others were thrown open to him. This is possible, though he was not dependent on their generosity. In Shakespeare's days each *bookseller's shop* was a sort of public library. Of these shops there was no lack in London, especially round about St. Paul's.

In conclusion one word about Shakespeare's sources and his originality. I look upon Shakespeare as the great architect, who gifted with a truly divine talent gave the materials their beautiful shape. The architect can never be made by the things. But he does not make the things either. The materials are given, not created by him. In so far, he is dependent on them. But more than this. His very conceptions and designs, however original they may be, are influenced by previously conceived plans and existent structures. In brief, originality is not creative production but novel combination.

[1] Hymen in As You Like It seems to be a later addition.—And who knows exactly how many *songs* are non-Shakespearean?

All this applies to Shakespeare. However great a genius he was, he was dependent on his 'materials'. He would never have become what he was, had he lived, say, in China from 1564 to 1616; or in England in the times of Hengist and Horsa. As a child of the western world he imbibes with his mother's milk certain ideas and modes of thought; as a child born in England in 1564 he is the inheritor of a language ready made, of all the literature of centuries, and of the culture and art of the Island. A heaven-born artist though he was, he had to learn much from his predecessors, many of them geniuses and men of the first rank. Their lessons he did not despise. He studied their writings carefully and diligently, until he became a truly *well-read* man.

A poet may at times be unconscious of the influences acting on him. But they are ever-present. And it is possible for a critic to know more about these influences than the poet himself.

The following quotations from Eckermann's Conversations with Goethe may serve as an apology for the present treatise:—

> People (said Goethe) are always talking about originality; but what does that mean? As soon as we are born, the world begins to act on us, and this goes on to the end. And, after all, what can we call our own except energy, strength, and will? If I could give an account of all that I owe to great predecessors and contemporaries, there would be but a small balance in my favour.[1]

Again, on the 17th February 1832 Goethe uttered the following weighty words very applicable, *mutatis mutandis*, to our investigation:—

> We then spoke further of Dumont; particularly of the memoirs which he wrote with reference to Mirabeau, and in which he reveals the various expedients (Hilfsquellen) which Mirabeau had contrived to employ, and also mentions by name many persons of talent, whom he had set in motion for his purposes, and with whose powers he had worked.
>
> "I know no more instructive book", said Goethe, "than these memoirs; by means of which we get an insight into the most secret recesses of that time, and by means of which the wonder Mirabeau becomes natural to us, while, at the same time, the hero loses nothing of his greatness. But

[1] I am quoting from the translation in Bohn's Standard Libr. 1874. The translation is not of superlative quality. I have made a few alterations. The above conversation is dated 1825, 12th May. — It is unfair to ask the question, who was the greater of the two, Shakespeare or Goethe. Their messages were quite different. The one is the exponent of the human heart, the other of nature at large. The one is *more* of the great artist and psychologist, the other of the intuitive advanced modern thinker.

now we have the latest critics of the French journals, who think a little differently on this point. These good folks think that the author of these memoirs wants to spoil their Mirabeau, because he unveils the secret of his superhuman activity, and allows other people a share in the great merit which, until now, the name of Mirabeau had monopolized.

"The French look upon their Mirabeau as their Hercules—and they are perfectly right. But they forget that even the Colossus consists of individual parts, and that even the Hercules of antiquity is a collective being—the idealized performer of his and others' deeds.

"But, in point of fact, we are all collective beings, do what we may. For how little have we, and are we, that we can strictly call our own property? We must all receive and learn both from those who were before us, and from those who are with us. Even the greatest genius would not go far if he willed to draw everything out of his own internal self. But many very simple-minded men do not comprehend that; and they grope in darkness for half a life, with their dreams of originality. I have known artists who boasted of having followed no master, but of being indebted to their own genius for everything. Fools! as if that were possible at all; and as if the world did not force itself upon them at every step, and make something of them in spite of their own stupidity. Yes, I maintain that if such an artist were only to survey the walls of this room, and cast only passing glances at the sketches of some great masters, with which they are hung, he would necessarily, if he had any genius at all, quit this place another and a higher man. And, indeed, what is there good in us, if it is not the power and the inclination to appropriate to ourselves the resources of the outward world, and to make them subservient to our higher ends. I may speak of myself, and may modestly say what I feel. It is true that, in my long life, I have done and achieved many things of which I might certainly boast. But to speak the honest truth, what had I that was properly my own, save the ability and the inclination to see and to hear, to distinguish and to select, and to enliven with some wit what I had seen and heard, and to reproduce it with some degree of skill. I by no means owe my works to my own wisdom alone, but to a thousand things and persons around me, who provided me with material. There were fools and sages, minds enlightened and narrow, childhood, youth, and mature age—all told me what they felt, what they thought, how they lived and worked, and what experiences they had gained; and I had nothing further to do than to put out my hand and reap what others had sown for me.

"It is, in fact, utter folly to ask whether a person has anything from himself, or whether he has it from others; whether he operates by himself, or operates by means of others. The main point is to have a great will, and ability and persistence to carry it out. All else is indifferent. Mirabeau was therefore perfectly right, when he made what use he could of the outer world and its forces. . . . This very peculiarity, that he understood how to act with others, and by others,—this was his genius—this was his originality—this was his greatness".

Lest we, in our eagerness to enquire into the sources, should overlook the all-importance of the architectonic Soul, the νοῦς which brings Harmony into Chaos, I quote the following passage.—Eckermann having stated that it were ridiculous to hunt up the sources of an author and then deny his originality, Goethe replied:

"That is very ridiculous; we might as well question a robust man about the oxen, sheep, and swine, which he has eaten, and which have given him strength.

"We are indeed born with faculties; but we owe our development to a thousand influences of the great world, from which we appropriate to ourselves what we can, and what is suitable to us. I owe much to the Greeks and French; I am infinitely indebted to Shakespeare, Sterne, and Goldsmith; but in pointing out these I have not exhausted the sources of my culture; the enquiry would proceed *ad infinitum* and were unnecessary. The chief thing is to have a Soul which loves the true, and which imbibes it wherever it is found.

"Besides, the world is now so old, so many eminent men have lived and thought for thousands of years, that there is little new yet to be discovered or said."[1]

The transcendent works of the great poet can never be said to be *explained* by the sources, he used. The same books were available to all his contemporaries, the same influences were at work everywhere,—why did the age not produce more than one Shakespeare? Why did his later contemporaries, with Shakespeare's example before their eyes, not write as great or greater works? Can all the sources we discover give the secret of those marvellous delineations by Shakespeare, of his fine poetic frenzy, of those 'skyey sentences,—aerolites,—'which seem to have fallen out of heaven'?[2]

In dealing with the problem of the sources of a poet, we constantly have to keep in mind that identity in thought in the productions of two writers need not necessarily imply dependence of one on the other. Neither a solitary parallelism nor a number of parallelisms prove anything, unless further corroborative evidence can be adduced:

---

[1] Date, Dec. 16, 1828.

[2] Emerson, Essay on Shakespeare.

"The world," said Goethe[1], "always remains the same; the conditions are repeated; one people lives, loves, and feels like another; why, then, should not one poet write like another? The situations of life are alike; why, then, should those of poems be unlike.

This we have not lost sight of.

Many of what are called Shakespeare's sources contain the mere embryos of his works. How despicable, for example, is 'The Famous Victories of Henry V.' in comparison with Shakespeare's Henry IV. and V. None the less, it is a production of perennial interest now, as showing the wonderful transformation which the subject received under the great master's hands. In other cases Shakespeare follows his source very closely. Nothing can be more interesting and instructive than an hour spent in Shakespeare's studio, where we can watch him actually at work upon his materials. We get into closer touch with him and we arrive at a better understanding. Would Goethe not have altered some of his judgments concerning Shakespeare, if he had known more of his environments and antecendents?

What were Shakespeare's *chief sources?*—The answer to this question which I have been frequently asked is: English dramatic works and the English Literature generally inclusive of the popular literary productions, Holinshed in special, Plutarch, the Bible, and Ovid.

'Shakespeare never will become popular', says one. I think, he will become more and more popular, not by study, not by learned comments and dissertations, but by good acting on the stage, which is the only true interpreter of his works.[2]

Go little book, to subtle world,
And show thy simple face,
And forward pass, and do not turn,
Again to my disgrace.

LONDON,
September, 1903.

---

[1] Jan. 18, 1825.

[2] It is only some months back, that I saw for the first time Twelfth Night acted, at Berlin. Now I understand the play; before I did not. But acting Shakespeare *well* is as hard as playing Beethoven well.

A SYNOPSIS

OF THE FOLLOWING ESSAY.

I. The Main Sources or Shakespeare's Plot-Books.

[N B.

— signifies that there is some probability of the former existence of a more direct source (*possibly* an early play) than the one mentioned.

* signifies: lost.

? signifies: more or less doubtful.]

Shakespeare's Works.	*Their Sources.*
Julius Cæsar	North's Plutarch.
Coriolanus	North's Plutarch.
Antony and Cleopatra	North's Plutarch.
King Richard II.	Holinshed.
Macbeth	Holinshed.
King Henry VI. Parts I., II., & III.	Holinshed and Hall.[1]
King Richard III.	Holinshed and Hall.[1]
King Henry VIII.	Holinshed and Foxe.
King Henry IV. Parts I. & II. King Henry V.	Holinshed and old play.
Cymbeline	Holinshed (for hist. framework); *old play(?) (for the love story, ultimately derived from Boccaccio).
King John	old play.
King Lear	old play.
The Taming of the Shrew	old play.
Measure for Measure	old play.
The Comedy of Errors	—Plautus.
Hamlet, Prince of Denmark	*old play.
The Merchant of Venice	*old play.
Timon of Athens	*old play, and North's Plutarch.
Twelfth Night	*old play(?), or Riche(?).

[1] Hall, probably as embodied in Grafton's Chronicle.

Shakespeare's Works.	*Their Sources.*
The Two Gentlemen of Verona	Montemayor transld. by Yong; or *old play.
Much Ado About Nothing	—Bandello.
Othello, the Moor of Venice	—Cinthio.
The Merry Wives of Windsor	—Italian tales.
All's Well that Ends Well	Boccaccio transld. by Painter.
As You Like It	Lodge's Rosalynde (Engl. romance).
The Winter's Tale	Greene's Pandosto (Engl. romance).
Romeo and Juliet	Arthur Brooke's poem.
Troilus and Cressida	Chaucer and Caxton.
Pericles (only in part by Shakespeare)	Gower and Twine.
Titus Andronicus	apparently not founded on any plot-source.
Love's Labour's Lost	do. „ do. „
A Midsummer-Night's Dream	do. „ do „
The Tempest	do. „ do. „
Poems:	
Venus and Adonis	Ovid.
The Rape of Lucrece	Livy and Ovid.
A Lover's Complaint	in the vein of Daniel's Rosamond.
Sonnets	English Sonnetists.
The Phoenix and the Turtle.	
The Passionate Pilgrim (contains some Shakespearean pieces.)	

II. Other Books and Literary Productions, reflected in Shakespeare's Works.

The Plays of Marlowe, Lyly, Kyd, Peele, Greene, Gascoigne, Preston, Alexander, etc. Mysteries and Moralities.

Poems of Marlowe, Daniel, Chaucer, Sidney, etc.

Prose Writings of Lyly, Sidney, Harsnet, Ralegh, Eden, etc.

School-books and Classical Authors. The Horn-book, ABC-book with the Catechism, Lily's Latin Grammar, Æsop's Fables, Mantuanus's Eclogues, Ovid, Cæsar, Seneca, Plautus (see above), Pliny, etc. Ovid he certainly knew both in the original and in the English translation; but Plutarch (see above) and Heliodorus in English dress alone.

French Authors. Montaigne in Florio's Translation, Ronsard, Rabelais.

As to *Spanish* and *Italian* authors comp. above.

Religious Books. The Bible (with the Apocrypha)—no work is more often quoted or alluded to by Shakespeare—, Book of Common Prayer, with the metrical Psalms.

Romantic Stories of Arthur, Guy, Bevis, etc.

Ballads of Robin Hood, Cophetua, Susanna, Jephthah, etc.

Many Songs; Roundelays; Popular Rhymes; Popular Tales; Proverbs; Folk-lore. Much of his knowledge and information Shakespeare, of course, derived from the lips of other men.—(Comp. also Contents.)

INTRODUCTION.

The first attempt at investigating the sources whence Shakespeare drew the plots of his dramas was made by Gerard Langbaine, who gave some notes on the originals of most of our poet's plays in his 'Momus Triumphans: or the Plagiaries of the English Stage', 1688,[1] and fuller notes in his scholarly work, 'An Account of the English Dramatick Poets', printed 1691. The section on 'William Shakespear', in the latter work, has been reprinted by the New Shakspere Society, Ser. IV, 3, pp. 318—331. Langbaine's remarks are of considerable interest to the student. The editors of Shakespeare of the eighteenth century were no doubt all indebted to his notes. In 1753—4 Mrs. Charlotte Lennox published a collection of alleged sources used by the poet, under the title of 'Shakespear illustrated or the Novels and Histories on which the Plays of Shakespear are founded, collected, and translated, etc.', 3 vols. This work was the precursor of 'Shakespeare's Library' edited by Collier in 1843, and re-edited by Hazlitt in 1875. The collection entitled 'Six old Plays on which Shakespeare founded [some of his plays]', published by Nichols in 1779 at the suggestion of Steevens, is the forerunner of Part II. of Collier-Hazlitt's 'Shakespeare's Library'. A work important for the study of the old drama was Robert Dodsley's 'Select Collection of Old English Plays', 12 vols., 1744. (The second edition prepared by Isaac Reed appeared in 1780, the third by J. P. Collier in 1825, the fourth comprehending 15 vols. by W. C. Hazlitt in 1874.)

The interest in Shakespeare had never ceased in England, but it was increased and deepened when the romantic movement set in towards the middle of the eighteenth century. Instead of the application

[1] Dict. of N. Biogr. gives the date 1687.

of classical standards to Shakespeare, as had been done of old, we now notice the clear and definite endeavour to understand and explain historically the great dramatist. The more distiguished Shakespearean scholars of the latter part of that century, to whom many valuable remarks on the literary influences acting on Shakespeare are due, are Steevens, Malone, Capell, Reed, Douce, etc.

In Germany, where the study of Shakespeare was taken up with enthusiasm, the following work was of importance for source enquiries: Karl Simrock, 'Die Quellen des Shakspeare', Bonn 1870; the first edition having appeared in 1831.[1]

The nineteenth century, the century of clubs and companies, of alliances and federal unions, has seen the birth (and also the death) of numerous Shakespeare Societies. The study of Shakespeare generally, of the sources of his works, and of the old drama has been furthered especially by the Old and New Shakespeare Societies, now extinct, in England, and by the Deutsche Sh. Gesellschaft, still vigorously alive. The best Shakespearean scholars have been members of these societies.

The literature which has sprung up around the name of Shakespeare has swelled to an enormous extent. Months are necessary for general orientation and years for a full command of the subject. People sometimes forget that it is better to read Shakespeare himself than works about him. And, alas, so much dilettantism is rife. Let us insist on the *ne multa sed multum.—Paucis opus est literis ad mentem bonam.* Our and future generations will only be able to glean stalks on the field of Shakespearean research. The splendid harvest has been gathered into the barns. But there is enough work still to be done. The sheaves must be thrashed out. The chaff, of which there is but too much, has to be winnowed from the grain, and the corn to be sifted and prepared for use.

It is now time that we should collect, utilise, and summarize the results of the labours of other men. Some work of this kind has been done; witness Alexander Schmidt's Lexicon, Sidney Lee's Life of Shakespeare, Halliwell-Phillipps' Outlines, Dowden's Primer. Mr. Furness is editing a new variorum edition. He will never finish it. But his son, who has re-edited his 'Macbeth', will carry it on. Malone's variorum edition was ably utilised by Delius in his stereotyped

[1] I ought to mention two essays of which I have availed myself, and which, I hope, I have now made superfluous, the one on 'Shakespeare's Library', the other on Shakspere's 'Classical Attainments', both of which appeared in 'Noctes Shaksperianae', published by the Winchester College Shaksp. Society, 1887.

edition of Shakespeare. We need another Delius to give us the gist of Furness's variorum edition. Other books are waiting to be written: A full Shakespeare-Bibliography; a scientific History of Elizabethan Literature; a carefully revised edition of Fleay's History of the Stage[1], which will honestly give its authorities, and state its guesses; an up to date edition of Simrock's 'Quellen'; and a revised edition of Collier-Hazlitt's 'Shakespeare's Library', which the German Shakespeare Society hopes to call into being before long. A working index to Arber's Transcript of the Stationers' Registers, and general indexes to the publications of the Shakespeare and Ballad Societies, are much needed.

[1] I hear that a work of this kind is in preparation.—I may also mention, that Dr. Furnivall is editing, and Prof. Liddel printing, "Shakespeare in Old Spelling". This is the work which scholars will quote for scientific purposes. I am still obliged to quote the text of the Globe edition.

CHAPTER 1.

SHAKESPEARE AND THE CLASSICS.

AN enquiry into Shakespeare's Knowledge and his sources has to give an answer to each of the following questions: Was Shakespeare acquainted with the classical authors? What education had he received? What modern languages did he know? What were his acquirements in English literature?

The question regarding the poet's education and learning has proved of remarkable attraction from the first. The learned Ben Jonson, in his commendatory verses prefixed to the First Folio, 1623, said, 'though thou hadst small Latin and less Greek',—a verdict which is not devoid of a smack of superciliousness. That Shakespeare was no match for Ben Jonson, as far as knowledge of the ancient classics is concerned, goes without saying. But that he had no need to be ashamed of his classical attainments, we shall soon see. For generations, however, it became a standing phrase that Shakespeare lacked Learning and Art.[1]

At a later date men like Upton, Whalley,[2] Pope, and Warburton were persuaded that Shakespeare was far more indebted to the ancients

[1] Compare Milton's 'native wood-notes wild' (L'Allegro, l. 134). We may add 'a quotation from the *Theatrum Poetarum* of Milton's nephew Edward Phillips, 'published in 1675. Milton had been dead a year; but he had trained Phillips 'and formed his tastes in poetry, and had probably helped him with hints for 'this very book. "In Tragedy", says Phillips of Shakespeare, "never any expressed '"a more lofty and tragic highth, never any represented nature more purely to '"the life; and where the polishments of art are most wanting, as probably his '"learning was not extraordinary, he pleases with a certain wild and native '"elegance.'" (Masson's ed. of Milton, vol. III, 1874, p. 377.)

[2] Peter Whalley, An Enquiry into the Learning of Shakespeare, 1748. An unsatisfactory and rambling booklet it is.

than had been commonly allowed. Against critics of this school, who claimed an amount of classical learning for the poet more than really belonged to him, Dr. Farmer levelled his heavy guns in his Essay on the Learning of Shakespeare (1767). But while refuting his opponents he fell into an error of the opposite extreme. In his essay, which, serviceable though it was, has been over-estimated by many people whose eyes were dazzled by its hyper-learnedness, he arrived at the following conclusion, which long remained a dogma of many Shakespearean scholars: 'He remembered perhaps enough of his *school-boy* 'learning to put the *Hig, hag, hog,* into the mouth of Sir Hugh Evans; 'and might pick up in the writers of the time, or the course of his 'conversation, a familiar phrase or two of French or Italian; but his '*studies* were most demonstratively confined to *nature* and *his own* '*language.*'

It was therefore consistent for men of this belief to make diligent search among English translations, mouldy chap-books, and dusty pamphlets. When, for example, no translation of Ovid's Fasti was discoverable, the Stationers' Registers were thumbed in the hope of finding traces of an old ballad of Lucrece, or Painter's free translation of Livy's version of the story was proclaimed as the source of Shakespeare's poem on the same subject.

Farmer's view was successfully contested by Dr. Maginn in Fraser's Magazine, Sept. 1839, and by Joseph Hunter, in his New Illustrations to Shakespeare, 1845, vol. II., pp. 313 ff. The enquiry into the poet's education and learning was further pursued from a different and better standpoint of view by Prof. Thos. S. Baynes, whose papers on 'What 'Shakespeare learnt at School' are well-known to the student of Shakespeare. They first appeared in Fraser's Magazine 1879—'80 and have since been published posthumously in Baynes's 'Shakespeare Studies', 1894.

Before entering on the subject we have set ourselves in the present chapter, I desire to make one short remark. This essay, as all essays are, more or less, is one-sided. It deals only with Shakespeare's book knowledge and with literary influences generally as visible in his works. The more important and powerful influences, however, on Shakespeare's young mind did not emanate from books. Nor was he a bookworm at any time, though he read a great deal in later life, digesting thoroughly and assimilating all he read. About Shakespeare the boy, let me quote Dr. Furnivall's golden words:[1]

[1] Introd. to the Leopold Shaksp., p. XII. An entertaining book of 256 pages on Shakespeare the Boy has been written by W. J. Rolfe, London, 1900.

'Shakspere, and his life as a Stratford lad, must be left to the fancy 'of every reader . . . . Taking the boy to be the father of the man, 'I see a square-built yet lithe and active fellow, with ruddy cheeks, 'hazel eyes, and auburn hair, as full of life as an egg is full of meat, 'impulsive, inquiring, sympathetic; up to any fun and daring; into 'scrapes, and out of them with a laugh; making love to all the girls; 'a favourite wherever he goes—even with the prigs and fools he 'mocks;—untroubled as yet with Hamlet doubts; but in many a 'quiet time communing with the beauty of earth and sky around him, 'with the thoughts of men of old in books;[1] throwing himself with 'all his heart into all he does.'[2]

But whatever the young genius with the best brains in England did in his hours of freedom, he certainly spent a great part of the day in the school-room. He received some mental training there. Of what nature was it? and what did he learn there? WHAT WERE HIS SCHOOL-BOOKS?

Young Will Shakespeare probably entered the Grammar School of Stratford in 1570 or 1571 at the age of six or seven years. In 1571 Shakespeare's father was chief alderman of the town. In 1568 he had been chosen bailiff (mayor) for a year. The boy was therefore of a very respectable Stratford family. There is some reason to suppose that he left school about 1578. The attendance was free of charge.[3]

[1] 'I don't press the books point', says Dr. F., 'except they were story-books, 'such as then existed.' Story books such as Captain Cox possessed. Robert Ashley ('an esquire's son', born 1565, at school in Southampton) "tells us that when a "boy he delighted in reading 'Bevis of Hampton', 'Guy of Warwick', 'Valentine "and Orson', 'Arthur and the Knights of the Round Table', and afterwards the "'Decameron of Boccacc' and the 'Heptameron of the Queen of Navarre.'" (Dict. Nat. Biogr., s. v. Rob. Ashley.)

[2] Speaking of Sh. when he first came to London, Dr. F. says: 'I believe 'in life and go as the essence of young Shakspere. He'd have wiped boots with 'a shoeclout, cleand a horse, commanded the channel-fleet, the army, or the nation, 'or written a sermon for any Romanist or Puritan, to say nothing of poems and 'plays for young nobles and the stage.' (Furn., *ut sup.*, p. XVI.)

[3] For information on Shakespeare's school work I am largely indebted to Prof. Thomas Spencer Baynes (*ut sup.*) and to Lupton's letter in the Athenaeum 1876, 7th Oct.; see Furnivall, Introd. to the Leopold Sh., pp. XI & CXXIV. Lupton's Life of Dean Colet contains some valuable notes. A highly interesting list of school books will be found in Arber's Transcript of the Stat. Registers, vol. III, 669—'70. The entry is dated 1620. Concerning the Horn-Book, see Tuer's History of the Horn-Book', 1897.—A list of Latin, Greek, French, Italian, and

The boy first had to learn reading by aid of the horn-book, which consisted of a slab of wood (or of other substance) in size usually rather less than 5 × 3 in., with a handle at one end. A printed sheet, containing the criss-cross, the alphabet, the vowels, some elements of spelling like ab, ba etc., and the Lord's Prayer, was glued down to the wood and covered by a thin plate of transparent horn. Here is a reproduction of 'a fine horn-book, indeed, in condition 'almost perfect', as shown in cut 131 of Mr. Tuer's Book. 'It was 'made in the days of Charles II.' Earlier examples of wooden horn-books are all damaged more or less. The horn-books of about 1570 were no doubt all in black letter.

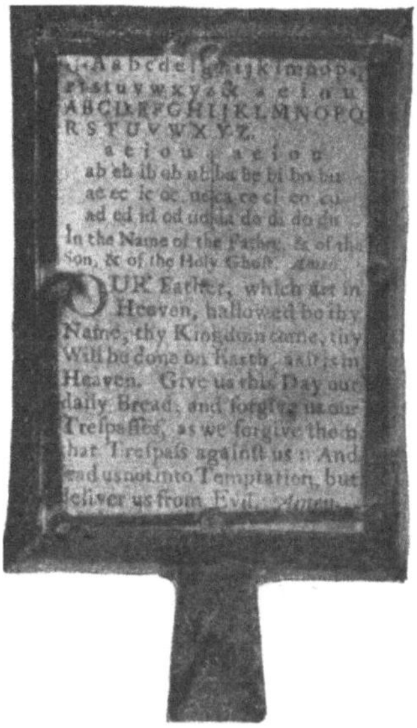

A Horn-Book (reduced size).

A first reading-book given to the young was the 'ABC book' which contained reading exercises and religious matters with the catechism.[1] The two other R's had to be learnt too, of course. Copy-books had been already introduced, to judge from Love's Labour's Lost, V, II, 42: "Fair as a text B in a copy-book". Counters, that is to say round pieces of metal, were characteristic aids of earlier

Spanish words in Shakespeare's works has been drawn up by A. Schmidt in the Appendix to his Sh.-Lexicon.—Stapfer's 'Shakespeare et l'antiquité' and Sarrazin's 'Shakespeare's Lehrjahre' deserve to be mentioned too.

[1] See below, p. 48.

days in arithmetical operations. Shakespeare alludes to them in several passages, notably in Wint. Tale, IV, II, 33 ff., where the clown confesses, "I cannot do't without counters".[1]

As soon as the boy had mastered the first rudiments, which were apparently taught by a pupil teacher or 'A-B-C-darius',[2] he was ready for the higher curriculum of the grammar-school.[3] What the curriculum of a school of a smaller town would consist of has been carefully studied by the late Prof. Thomas Spencer Baynes, whose guidance we are safe in following. I shall quote his own words, inserting within square brackets a few remarks which are in full harmony with other passages in Baynes's paper and with Lupton and Furnivall's list. 'From these various sources, contemporary and quasi-contemporary, 'we may form a trustworthy general estimate of Shakespeare's course 'of instruction during his school days. At the time, as we have seen, 'boys usually went to the grammar school about six or at latest seven 'years of age, and entered at once upon the Accidence. In his first 'year,[4] therefore, Shakespeare would be occupied with the Accidence 'and grammar [namely, Lily's Grammar]. In his second year, with 'the elements of grammar, he would read some manual of short 'phrases and familiar dialogues, and these committed to memory would 'be colloquially employed in the work of the school [some manual 'like Sententiae Pueriles, Pueriles Confabulatiunculae, Corderius's 'Colloquies]; in his third year, if not before, he would take up Cato's 'Maxims[5] and Æsop's Fables; in his fourth, while continuing the

[1] Comp. too Troilus, II, II, 28; As You Like It, II, VII, 63; Caesar, IV, III, 80; Cymbeline, V, IV, 174; Othello, I, I, 31.

[2] Shakespeare's parents could not have taught him writing, as they could not even sign their names. To think of a modern mayor who could not write his name!—The second schoolmaster at Chigwell, Essex, had to be 'skilful in 'cyphering and casting of accounts' and to 'write fair secretary and Roman hands'. (Lupton, Athen., *ut sup.*)

[3] In Rotherham, e. g., 'the custom was to enter boyes to the Schoole one by ,one, *as they were fit for Accidents.*' (Fras. Mag., 1879, p. 618.)

[4] The word 'year' used in the following sentences should not be taken too literally. At any rate, Horace, Juvenal, Persius, and Seneca are mentioned by Hoole and Brinsley (Baynes's authorities) as works read in the highest class, the boys of which were about 15 years of age, or in their 8th school-year. If we suppose that Shakespeare left school at an earlier date (for which supposition, however there is no cogent reason, seeing that his father had nothing to pay for the instruction of his son) he could scarcely have been very familiar with these authors.

[5] On Cato's Disticha de Moribus see Warton, Hist. of Engl. Poetry, III, pp. 133 ff. On the Sententiae Pueriles see below, p. 47.

'Fables, he would read the Eclogues of Mantuanus, parts of Ovid, 'some of Cicero's Epistles, and probably one of his shorter treatises; 'in his fifth year he would continue the reading of Ovid's Meta-'morphoses, with parts of Virgil and Terence; and in the sixth Horace, 'Plautus, and probably part of Juvenal and Persius, with some of 'Cicero's Orations and Seneca's Tragedies. In going through such a 'course, unless the teaching at Stratford was exceptionally inefficient, 'the boy must have made some progress in several of these authors, 'and acquired sufficient knowledge of the language to read fairly well 'at sight the more popular poets and prose writers such as Ovid and 'Cicero. [The Greek grammar, if any, in use at Stratford, would most likely be Clenard's 'Institutiones, absolutissimae in Graecam linguam.'] 'The masters of the school during the time Shakespeare attended it 'would seem, however, to have been at least of average attainments 'and ability as they rapidly gained promotion. No fewer than three 'held the post during the decade from 1570 to 1580. In the first 'two years Walter Roche, for the next five, the most important in 'Shakespeare's school history, Thomas Hunt, and during the last three 'years Thomas Jenkins were headmasters in the school.' Of geography, history, etc., probably a smattering was given, but I know no authority on the point. The chief subject of a grammar-school was Latin, and the principal book the Latin grammar. Hence the name 'grammar-school'.

Shakespeare probably learnt singing as a member of the Stratford church choir. The Bible and the Common Prayer-Book are very familiar to him. (See a separate chapter on this subject.)

On Baynes's essay Dr. Furnivall gives the following comment: 'He makes out a fair case'. But, adds Dr. F., 'what would Stratford 'provincials do, in Shakspere's day, with the large doses of Latin that 'the profest curriculums of larger town schools provided?' Dr. Furnivall does not appear to me to hit the nail on the head. Prof. Baynes continually refers to towns, such as Rotherham and Ashby-de-la-Zouche, which were certainly not large towns. But after all, Dr. Furnivall's list of Shakespeare's school-books (Leopold Sh., p. XI) does not materially differ from Baynes's list.[1]

[1] While the present chapter is passing through the press, the first part of Mr. J. C. Collins's article "Had Shakespeare read the Greek Tragedies?" is appearing in the Fortnightly Review (April 1., 1903). Collins refers us to the curriculum drawn up for the Ipswich Grammar School in 1528, where the following authors and books were prescribed: Lily's Grammar, Æsop, Terence, Virgil, Cicero, Sallust, Cæsar, Horace, Ovid, Donatus's Commentaries.

What traces of the above mentioned school-books do we find in Shakespeare's works?

Beginning with the inevitable

HORN-BOOK[1]

with its 'ba', consonants, and the five vowels, we have an interesting reference to it in Love's Labour's Lost, V, i, 47 ff.:

Armado [To Holofernes]: Monsieur, are you not lettered?
Moth. Yes, yes; he teaches boys *the horn-book.* What is a, b, spelt backward, with the horn on his head?
Hol. Ba, pueritia, with a horn added.
Moth. Ba, most silly sheep with a horn. You hear his learning.
Hol. Quis, quis, thou consonant?
Moth. The third of the five vowels, if you repeat them: or the fifth, if I.
Hol. I will repeat them,—a, e, i,—
Moth. The sheep: the other two concludes it.—o, u.

The alphabet, as we have seen, was preceded by a cross. Hence it was called Christ-cross-row, or criss-cross-row, or simply cross-row. Compare Richard III., Act I, i, 54:

He hearkens after prophecies and dreams;
And from the *cross-row* plucks the letter G,
And says a wizard told him that by G
His issue disinherited should be.

THE ABC BOOK WITH THE CATECHISM[2]

is referred to in King John, I, i, 192 ff.

Why then I suck my teeth and *catechize*
My picked man of countries: 'My dear sir',
Thus, leaning on mine elbow, I begin,
'I shall beseech you'—that is question now;
And then comes answer like an *Absey book*:
'O sir', says answer, 'at your best command;
'At your employment; at your service, sir:'
'No, sir', says question, 'I, sweet sir, at yours:'
And so, ere answer knows what question would,
. . . . It draws toward supper in conclusion so.

Compare, too, The Two Gentlemen, II, i, 23:

to sigh, like a school-boy that had lost his ABC.

But this may refer to the horn-book.

[1] Comp. *ante*, p. 9.
[2] See pp. 9 and 48.

LILY'S LATIN GRAMMAR.

William Lily (1468?—1522) was one of the earliest Greek scholars in England, a friend of Thomas More, Erasmus, and John Colet, who appointed him the first high master of St. Paul's School. About 1509 Colet wrote an Accidence of the Latin language ('Coleti Aeditio' with the rules in English), to which was added a short Syntax by Lily, also in English. About 1540 this Grammar underwent a thorough revision. A copy of the year 1568, when Shakespeare was four years old, bears the following title; 'A Shorte Introduction of Grammar 'generally to be used: compiled and set forth for the bringing up of 'all those that intende to attaine the knowledge of the Latine tongue'.[1] The title of the copy of the year 1577, in the Britsh Museum, is the same, excepting some differences of spelling.[2] From this edition I shall quote. I have also compared the text of earlier and later editions. The 1577 copy is partly in black letter and partly in Roman type. After a few pages of preliminaries, begins the Accidence or 'Introduction 'of the eyght partes of Latine Speache', which answers to Colet's 'Aeditio'. On folio 19. commences the Syntax ('The Concordes' and 'The Construction of the eyght partes of speache') which ends on leaf 27. This is based on Lily's Syntax added to Colet's Accidence. After some pages of Latin precepts and prayers we come to the Second Part, the title page of which is wanting in our copy. From other copies of the Grammar we know it ran thus; 'Brevissima Institutio seu 'Ratio Grammatices cognoscendae', etc. This part, to some degree a complete grammar of itself, is written entirely in Latin, while part 1. is in English. Nothing corresponding to this second part is found in the early editions of Colet and Lily's Grammar before 1540. This 'Brevissima Institutio' is compiled from Lily's 'de generibus nominum, ac verborum praeteritis et supinis regulae', to which T. Robertson made additions; from Lily's Syntax, with rules in Latin, entitled 'Absolutissimus de octo orationis partium Constructione libellus', a work entirely different from the Syntax referred to above, and in the composition of which Erasmus had a hand; and from other writings.[3]

The further history of Lily's Grammar does not concern us here. Suffice it to say that it long remained the national grammar of England, Having passed through various phases it finally developed into the Eton Latin Grammar of to-day.

[1] Title given by Halliwell-Phillipps, in Notes and Queries, 6th Ser., II, p. 462.

[2] Compare *post*, page 47, note.

[3] Compare John Ward's Preface to Lily's Grammar, published in 1752.

Shakespeare's acquaintance with Lily's Grammar, commonly known as the Accidence, is satisfactorily proved by the catechetical scene in The Merry Wives of Windsor, Act IV., sc. 1. Sir Hugh Evans asks the boy, William, 'some questions in his *accidence*'. The answer to Evans's query, 'how many numbers is in nouns?' will be found on the first page of the grammar proper: 'In Nounes be 'two Numbers, the Singular, and the Plurall. The Singular Number 'speaketh of one: as *Lapis*, a stone. The plurall number speaketh 'of mo than one; as *Lapides*, Stones'. Compare The Merry Wives, ll. 32:

Evans . . . What is '*lapis*', William?
Will. A stone.
Evans. And what is 'a stone', William?
Will. A pebble.
Evans. No, it is '*lapis*': I pray you, remember in your prain.
Will. *Lapis.*

Again, ll. 26—30:

Evans . . . What is 'fair', William?
Will. *Pulcher.*
Mrs. Quickly. Polecats! there are fairer things than polecats, sure—

refers to the same page, where '*Bonus*, Good; *Pulcher*, Fayre' are given as instances of adjectives.

On page 2. of Lily's Grammar we read:

Articles.

Articles are borowed of the Pronoune, and be thus declined.

Singulariter.		*Pluraliter.*	
Nominativo	*hic, hæc, hoc.*	*Nominativo*	*hi, hæ, hæc.*
Genitivo	*huius.*	*Genitivo*	*horum, harum, horū.*
Dativo	*huic.*	*Dativo*	*his.*
Accusativo	*hunc, hanc, hoc.*	*Accusativo*	*hos, has, hæc.*
Vocavito	*caret.*	*Vocativo*	*caret.*
Ablativo	*hoc, hac, hoc.*	*Ablativo*	*his.*

Compare with this The Merry Wives, *ut sup.*, ll. 39 ff: —

Evans . . . What is he, William, that does lend articles?
Will. *Articles are borrowed of the pronoun, and be thus declined, Singulariter, nominativo, hic, haec, hoc.*
Evans. *Nominativo, hig, hag, hog*; pray you, mark: *genitivo, hujus.* Well, what is your accusative case?
Will. *Accusativo, hinc.*
Evans. I pray you, have your remembrance, child: *accusativo, hung, hang, hog.*
Quick. 'Hang-hog' is Latin for bacon, I warrant you.

Evans. Leave your prabbles, 'oman. What is the focative case William?
Will. *O,—vocativo, O.*
Evans. Remember, William; focative is *caret.*
Quick. And that's a good root.
Evans. 'Oman, forbear.
Mrs. Page. Peace!
Evans. What is your genitive case plural, William?
Will. Genitive case!
Evans. Ay.
Will. Genitive.—*horum, harum, horum.*
Quick. Vengeance of Jenny's case! fie on her! never name her, child, if she be a whore.
Evans. For shame, 'oman. etc.

William's hesitative answer, "*O,—vocativo, O*", finds its explanation on page 2. of the Grammar, where we read: 'The Vocative 'case is knowne by calling or speaking to: as *O magister*, O mayster'. Thus we find in the paradigms: '*Vocativo ô musa*'; '*Vocavito ô lapis*', and the like.—Some pages further we have the declension of the pronoun: *qui, quæ, quod,* referred to by Evans, ll. 76—81.

In 1. Henry IV., Act II, I, 104, there is the following quotation from the Grammar (p. 1.):

homo, is a common name to all men.

On folio 19. there is a section on Interjections, of which the grammar gives examples like the following: 'Some are of myrth: as '*Evax, vah.* Some are of . . . Laughing: as *Ha ha he* . . . Calling: 'as *Eho, oh, io*', etc.—Compare Much Ado Ab. Nothing, Act IV., I, 22:

How now! interjections? Why, then, some be of laughing, as, ah, ha, he!

The phrases, *Diluculo surgere, saluberrimum est,* and *Vir sapit, qui pauca loquitur*, to be found on leaf 20. of Lily's Grammar, are referred to in Twelfth Night (II, III, 2) and in Love's Labour's Lost (IV, II, 82) respectively.

The line, *Redime te captum quam queas minimo*, in The Taming of the Shrew (I, I, 167) is not taken from Terence direct, but from Lily (see Part. II., *Abl. post verb.*), where the quotation is given in the *altered* form which we find in Shakespeare,—the original words in Terence's Eunuchus (I, I, 30) being:

[1] This witty reference to the Grammar may have been suggested by George Lyly's Endymion, where we find the following allusive passage:—'an interjection, 'where of some are of mourning, as *eho! vah!*' Cp. also 'Mother Bombie', III, II: 'interjections like winde, as eho, ho, to'.

Quid agas? nisi ut te redimas captum quam queas Minimo.[1]

Novi hominem tanquam te (Love's Labour's Lost, V, I, 10) and the phrase *ad unguem* (ibid., l. 84) may have also been taken from the Grammar (cp. Part II., Syntax of Adv^{s}; and folio 28., verso).

Lastly, in Titus Andronicus, IV, II, 20—3 we find quoted the lines of Horace:

Integer vitae, scelerisque purus,
Non eget Mauri jaculis, nec arcu.

Chiron, on hearing them, observes: 'O, 'tis a verse in Horace; I know 'it well: *I read it in the grammar long ago.*'—The couplet stands twice in Lily's Grammar: on leaf 23., as an instance of the Ablative case, and in Part II., under the head of '*De generibus carminum*, where Horace is named.

From which manual Shakespeare learnt his vocabularies and elementary phrases, I cannot say with certainty. But we find ample illustrations of Latin exercises and dialogues, such as would be practised at school, in Love's Labour's Lost, IV, II; and V, I. The following passages will serve as examples:—

Holofernes: The deer was, as you know, *sanguis, in blood;* ripe as the pomewater, who now hangeth like a jewel in the ear of *caelo, the sky, the welkin, the heaven;* and anon falleth like a crab on the face of *terra, the soil, the land, the earth.*

(Love's L. L., IV, II, 3—7.)

Again,

Nathaniel: Laus Deo, bone, intelligo.
Hol. Bone?—bone for *bene:*[2] Priscian a little scratched, 'twill serve.
Nath. Videsne quis venit?
Hol. Video, et gaudeo.

(Love's L. L., V, I, 30—34.)

Compare, too, 'The Taming of the Shrew', Act III, sc. I.

[1] Quoted from two Elizabethan editions of Terence. The line is cited and Englished thus by Udall in his 'Floures for Latine speakyng selected and gathered oute of Terence, and the same translated into englyshe' (1560): *Redimas te captum quàm queas minimo,* Redeeme or raunsome thy selfe, beynge taken prisoner, as good chepe [as cheaply] as thou maiest, or, if you be in any daunger, come out agayne as well as you maie.—Udall seems to have copied the sentence with the translation from the '*Bibliotheca Eliotae.*'

[2] This line is corrupt in the Folio, which reads: . . . '*bene intelligo. Peda. ,Bome boon for boon prescian,* a little scratcht'. I have given Theobald's emendation. The Globe edition reads: . . . 'bene intelligo. Bon, bon, fort bon!'

ÆSOP'S FABLES.

Gloucester's hint, in 3. Henry VI., Act V, v, 23, that the masculine queen, Margaret, should have always "worn the petticoat, And "ne'er have stol'n the breech from Lancaster", is met by the young Prince with the following caustic retort, containing an allusion to Gloucester's figure crooked like that of Æsop:

> Let *Æsop fable* in a winter's night;
> His currish riddles sort not with this place.

Strange to say, Henry Green[1] infers from this passage that Shakespeare had a low estimate of Æsop's fables. But the expression "His currish "[=malicious] riddles" in no wise warrants this inference. The words must be taken *cum grano salis* as referring to what Gloucester had just remarked, and not as derogatory from Æsop. The apt illustrations which Shakespeare drew from the famous fables leave no doubt that the poet had no mean opinion of them.

1. *The fable of the Countryman and a Snake*[2] is alluded to in 2. Henry VI., Act III, I, 343:

> I fear me you but warm the starved snake,
> Who, cherish'd in your breasts, will sting your hearts.

Compare, too, Richard II., Act III, II, 129—131:

> *K. Rich.* O villains, vipers, damn'd without redemption! . . . .
> Snakes, in my heart-blood warm'd that sting my heart!

And Act V, III, 57:

> Forget to pity him, lest thy pity prove
> A serpent that will sting thee to the heart.

2. *The Crow and the Borrowed Feathers* is alluded to by Shakespeare (who, by the way, was himself once called 'an upstart crow,

[1] Shakespeare and the Emblem Writers, 1870, p. 302.

[2] *Agricola & Anguis.* Repertum anguem frigore penè mortuum agricola misericordia motus, fovere sinu, & subter alas recondere. Anguis recreatus calore, vires recepit, ac confirmatus, agricolae, pro merito ipsius summo, letale vulnus inflixit.—Fabula demonstrat eam mercedem, quam rependere pro beneficiis mali consuevere.

I quote this from an edition of Æsop published in 1592, with forewords by Philippus Melanchthon, with the following title: Fabellae Aesopicae quaedam, notiores, et *in scholis usitatae* . . . a J. Camerario. 1592. Lipsiae. The fable is of course, also contained in other editions of Æsop of the sixteenth century, some of which I have examined.

'beautified with our feathers', by the dying Greene) in 2. Henry VI., Act III, I, 69ff.

King . . . Our Kinsman Gloucester is as innocent
From meaning treason to our royal person
As is the sucking lamb or harmless dove . . .
Queen . . . Seems he a dove? *his feathers are but borrow'd.*
For he's disposed as the hateful raven:
Is he a lamb? etc. (cf. below.)

And again in Timon of Athens, II, I, 28:

I do fear,
When every feather sticks in his own wing,
Lord Timon will be left a naked gull,[1]
Which flashes now a phoenix.

The allusion is not very plain if we think of the fable in the version with which we are probably most familiar. Here a jackdaw assumes the feathers of a peacock and is stripped bare by the birds it imitated. The allusion in Timon, however, becomes more pertinent if we compare the fable as it is told, for example, in the Latin Æsop (p. 17) referred to above.

De Cornice superbiente aliarum avium pennis. Cornicula collectas pennas de reliquis avibus sibi commodaverat, & superba varietate illa, reliquas omnes prae se aviculas contemnebat. Tum fortè hirundo notatā suā pennā, advolans illam aufert, quo facto & reliquae postea aves quaeq. suam ademere cornici: ita illa risum movit omnibus, furtivis nudata coloribus, ut ait Horatius.—Significat fabula, commendicatam speciem neq. diu durare, & perlevi momento dissolvi.

3. *The Ass in a Lion's Skin.* This fable we find referred to in King John, II, I, 139—146:

Bastard . . . *I'll smoke your skin-coat, an I catch you right* . . .
Blanch. O, well did he become that *lion's robe*
That did disrobe the lion of that robe!
Bast. It lies as sightly on the back of him
As great Alcides' shows upon an ass:
But, ass, I'll take that burthen from your back,
Or lay on that shall make your shoulders crack.

4. *The Wolf in a Sheep's Skin*[2] seems to be referred to in 2. Henry VI., Act III, I, 77—9:

[1] Gull = unfledged bird.

[2] *Lupus.* Induerat pellem ovis lupus, atq. cum ita ignoraretur, aliquantisper impunè in gregem fuit grassatus. Sed pastor mox animadversa fraude, necatum

Is he a lamb? His skin is surely lent him,
For he's inclined as is the ravenous wolf.
Wo cannnot steal a shape that means deceit?

But perhaps it is better to suppose that Shakespeare had in mind Matthew VII, 15.

5. *The Fox and the Grapes.* This well-known fable is alluded to in All's Well, II, I, 71 ff:

Lafeu . . . Will you be cured of your infirmity?
King. No.
Laf. O, will you eat no grapes, my royal fox?
Yes, but you will my noble grapes, an if
My royal fox could reach them.

6. *The Hunter and the Bear*, is probably the fable Shakespeare had in mind, when he wrote Henry V., Act IV, III, 91—94:

Bid them achieve me and then sell my bones.
Good God! why should they mock poor fellows thus?
The man that once did sell the lion's skin
While the beast lived, was killed with hunting him.

7. *The Oak and the Reed.* Of this fable, relating the overthrow of the oak which resisted the tempest, while the yielding reed (or the willow, according to other versions) received no harm, there may be a possible reminiscence, as Green suggests, in Cymbeline, VI, II, 267:

To thee the reed is as the oak—

and in Love's Lab. Lost, IV, II, 112:

Those thoughts to me were oaks, to thee like osiers bow'd.—

It is hardly necessary to refer to such general allusions as:

A lion and a king of beasts. (Rich. II., Act V, I, 34).

And to passages like,

thou hast entertain'd
A fox to be the shepherd of thy lambs (Gentlemen, IV, IV, 96),

or,

The fox barks not when he would steal the lamb
(2. Henry VI., Act III, I, 55),

we might easily cite illustrative passages from Æsop. The Greek fable writer, however, makes wolves, not foxes, the dangerous enemies of the lambs.

hunc de arbore suspendit. Hoc qui pelle decipiebantur admirantibus: Pellis quidem est, pastor inquit, ovis, sed sub hac lupus latebat.—Habitus et vultus indicia non habenda pro certis, fabula docet: ideoq. facta & rem spectari oportere.

From the above remarks it is clear that Shakespeare was familiar with some Æsopian fables. I regret to say that I could not meet with a Latin edition of them printed in England in the sixteenth century, excepting the editions of 1502 and 1503, in the British Museum.

MANTUANUS.

The *Bucolica* of Battisto Spagnuoli, a Carmelite monk (d. 1516), called Mantuanus after his birthplace Mantua, enjoyed much popularity, and was established as a text-book in many schools both in England and on the Continent. The opening words of the first eclogue are quoted by Holofernes in Love's Labour's Lost, IV, II, 95:

> Fauste, precor, gelida quando pecus omne sub umbra Ruminat,—and so forth. Ah, good old Mantuan! I may speak of thee as the traveller doth of Venice;
>
> Venetia, Venetia,
> Chi non ti vede non ti pretia.
>
> Old Mantuan, old Mantuan! Who understandeth thee not, loves thee not.

CÆSAR.

The following passage in 2. Henry VI., Act IV, VII, 65—68:

> Kent, in the Commentaries Cæsar writ,
> Is term'd the civil'st place of all this isle:
> Sweet is the country, because full of riches;
> The people liberal, valiant, active, wealthy—

seems to contain a reminiscence of an early school book, the Commentaries or *De Bello Gallico.* In the fourteenth chapter of the Fifth Book, which gives an account of the second Invasion of Britain by Cæsar together with a short description of the Island, we find the following sentence, obviously alluded to:

> Ex his omnibus longe sunt humanissimi, qui Cantium incolunt.[1]

To the Commentaries we find another reference in Richard III., Act III, I, 84—

> That Julius Cæsar was a famous man;
> With what his valour did enrich his wit,
> *His wit set down to make his valour live;*
> Death makes no conquest of this conqueror;
> For now he lives in fame, though not in life.

[1] Lyly, in his Euphues (Arber, p. 247), repeats Cæsar's remark: 'Of al the 'Inhabitants of this Isle, the Kentish men are most civilest'. But the author of 2. Henry VI. *knows* that this is from *Cæsar's Commentaries.*

CICERO.

I) In 2. Henry VI, Act IV, I, 108, there is a reference to "Bargulus "the strong Illyrian pirate",[1] of whom Shakespeare may have read in Cicero's '*De Officiis*': 'Bargulus, Illyricus latro, de quo est apud 'Theopompum magnas opes habuit'. Malone, in his Variorum Edition, II, 104 n., points out that this book was much read in schools.

II) In Titus Andronicus, IV, I, 12—14, Titus says of Lavinia:

Ah, boy, Cornelia never with more care
Read to her sons than she hath read to thee
Sweet poetry and *Tully's Orator.*

The title, at least, of Cicero's *De Oratore* was therefore familiar to the author of Titus Andronicus. I need scarcely remark that 'Tully', and not 'Cicero', was the usual name by which the Roman author was known in Shakespeare's days.

OVID.

It is my purpose to show that Ovid,[2] a favourite author with Shakespeare, was known to him both in the original and in the English translation, and to supply further evidence of his familiarity with the Roman poet.

I) OVID IN THE ORIGINAL.

On the title-page of 'Venus and Adonis' Shakespeare put as a proud motto the following couplet from Ovid's *Amorum Lib.* I., Eleg. XV, 35:

Vilia miretur vulgus; mihi flavus Apollo
Pocula Castalia plena ministret aqua.

From '*Metamorphoses*' (I, 150) is derived:

Terras Astræa reliquit. (Titus, VI, III, 4).

From *Heroides* (II, 66):

Di faciant, laudis summa sit ista tuae!
(3. Henry VI., Act I, III, 48).

From *Heroid.* (I, 33):

Hic ibat Simois; hic est Sigeia tellus;
Hic steterat Priami regia celsa senis. (Shrew, III, I, 28).

---

[1] The Quartos read: "mighty Abradas, the great Macedonian pirate".

[2] In the Bodleian Library there is a copy of Ovid's Metamorphoses in the Latin text, printed by Aldus, Venice, 1502. On the title-page is the signature W.m Sh.r. (cp. Shakesp. Jahrb. XVI, 367). It is not certain whether the autograph is genuine.

Baynes, Fraser's Magaz., 1880, pp. 101—2, has shown that the name 'Titania' for the fairy queen in Mids. N. Dream gives evidence of Shakespeare's intimate knowledge of Ovid in the original, for the name is not to be found in Golding's translation.[1]

A passage in King John, V, vii, 25—27:

For you are born
To set a form upon that indigest
Which he hath left so shapeless and so rude.—

reminds us of the following words, referring to Chaos, in *Metam.* I, 7:

rudis indigestaque moles,—

for which Golding has:

a huge rude heape, and nothing els but even.

Shakespeare probably knew in the original Ovid's *Ars Amatoria*, of which there appears to have been no early translation. It is read by Lucentio in The Taming of the Shrew (IV, ii, 8):

I read that I profess, *the Art to Love.*—

Compare, in conclusion, 'Romeo' II, ii, 92:

At lovers' perjuries
They say Jove laughs,—

and *Ars Am.* I, 633:

Jupiter . . . perjuria ridet amantum.

ii) OVID IN THE TRANSLATION.

Of Ovid there existed the following translations in Shakespeare's days:

The Metamorphoses translated by Golding (first complete edition, 1567); the Elegies (Amores) by Marlowe (printed *circa* 1597); the Heroical Epistles by Turbervile (1567); the three first Bookes of Ovid de Tristibus by Thomas Churchyard, 1580. Besides these, other portions of Ovid were translated. (Cp. Malone I, 381, and Warton's Hist. of Poetry IV, 293ff.)

That Shakespeare had in his mind a passage from Golding's Ovid when he penned Prospero's incantation in The Tempest, V, i, 33, was

[1] Malone, Var. Ed. II, 337, asserts that Oberon and Titania had been introduced in a dramatic entertainment exhibited before Queen Elizabeth in 1591, when she was at Elvetham in Hampshire. This statement is not correct. 'Auberon' is, indeed, referred to there as the Fairy King; but the 'Fayery Quene' calls herself ,Aureola', not Titania! See Nichols's Progr. of Q. Eliz., III, 101ff. (As to Oberon see *post*, s. v., Charlemagne Romances.)

pointed out by Farmer. The whole passage will be found in Malone's Variorum Edition, XV, 160, to which I refer the reader. The proof of Shakespeare having used Golding rests especially upon the comparison of the following verses:

Temp: "*Ye elves of hills, brooks, standing lakes* and groves!"
Ovid (Met. VII, 197): 'Auraeque et venti, montesque amnesque lacusque,
'Dique omnes nemorum, dique omnes noctis adeste'.
Golding: 'Ye aires and winds, *yee elves of hills*, of *brooks*, of woods alone,
'Of *standing lakes*, and of the night approach ye everichone.'

Not quite so striking is the verbal agreement between Shakespeare's 'Venus and Adonis', where he says of the boar,—

v. 619: On his bow-back he hath a *battle set*
Of *bristly pikes*, that ever threat his foes;
His eyes, like glow-worms, shine when he doth fret.
v. 625: His *brawny* sides, with hairy *bristles* arm'd,
Are better proof than thy spear's point can enter, etc.,—

and Golding's translation of *Metam.* VIII, 284f., where the Calydonian boar is described as follows:

His eyes did glister blood and fire, right dreadful was to see
His *brawned back*; right dreadful was his hair which grew as thick
With pricking points as one of them could well by other stick:
And like a front of armed *pikes*, *set* close in *battle ray*,
The sturdy *bristles* on his back stood staring up alway.

In the original we read:

Sanguine, et igne micant oculi. riget horrida cervix;
Et setae densis similes hastilibus horrent;
Stantq velut vallum, velut alta hastilia setae.[1]

THE ACTÆON MYTH AND GOLDING.

In Mids. N. Dream, IV, I, 107ff., Theseus orders the hounds to be uncoupled in order that Hippolyta may hear their music. The queen then relates:

I was with Hercules and *Cadmus* once,
When in a wood of *Crete* they bay'd the bear
With hounds of *Sparta:* never did I hear
Such gallant chiding; for, besides the groves,
The skies, the fountains, every region near
Seem'd all one mutual cry: I never heard
So musical a discord, such sweet thunder.

[1] I quote these lines from Aldus's edition, 1502.

Theseus: My hounds are bred out of the *Spartan* kind,
So flew'd, so sanded . . .

In illustration of these verses the commentators quote from Ovid's narrative of Actæon, grandson of Cadmus (Cadmus, by the way, occurs also in The Knight's Tale by Chaucer), in Golding's translation.

In the original text we read:

(Met. III, 208) *Gnosius* Ichnobates, *Spartana* gente Melampus.
(Met. III, 223) Et patre Dictæo, sed matre Laconide nati
Labros et Agriodus, et acutae vocis Hylactor,
Quosque referre mora est.

Golding, referring to the two dogs of v. 208, says:

'The latter was a hound of *Crete*, the other was of *Spart*.'

In the Latin original of the Actæon narrative the name Crete (or Creticus, etc.) nowhere occurs. The verses 223—4, quoted above, are translated very *freely* by Golding:

. . . with other twaine that had a sire of *Crete*,
And dam of *Sparta:* tone of them cald iolliboy a great
And large *flewd* hound: the tother chorle who ever gnarring went,
And *ring-wood* with a shrill lowd mouth the which he freely spent,
With divers mo, whose names to tell it were but losse of time,

These coincidences, noteworty though they are, can scarcely be looked upon as singularly striking. There is yet another passage in The Merry Wives which, taken in connection with the verses just quoted from Golding, calls for a remark. In Act II, I, 121—2 of this drama, Pistol is made to say,

Prevent, or go thou
Like Sir *Actæon* he, with *Ringwood* at thy heels.

Here, it might be supposed, we have a clear proof of Shakespeare having used Golding. But it is probably not so. If the dramatist had referred to Golding the allusion would appear rather pointless. Who would have understood it? And how should we suppose Shakespeare to have hit upon *Ringwood*, one of a large number of hounds recorded by Golding. The truth seems to be, the connexion with the translator of Ovid is of an indirect kind. In a *song* entitled 'Mad Tom', or 'New Mad Tom of Bedlam', there occur the following lines:

Poor naked Tom is very dry—
A little drink for charity!
Hark! I hear *Actæon's* hounds!
The huntsmen whoop and hallowe;
'*Ringwood*, Royster, Bowman, Jowler',
All the chase now follow.

Now the allusion seems more intelligible. Here we have a *song* in which Ringwood is the first named dog at the heels of Actæon, while in Golding's list of hounds he is in no wise prominent. The song is to be found in Percy's Reliques; Ballad Soc., Roxb. Ball., 2, p. 259; and Chappell's Old Pop. Mus. The names of the hounds (excepting Jowler) are all in Golding's version. There can be no doubt that the writer of the song had an eye to this translation of Ovid. The exact date of the poem is not known. 'One of the ballads, directed 'to be sung to the tune of *Mad Tom*, and which is in the same measure 'as the song [referred to] above, adds to the direction—"as it was 'lately sung at the Curtain, Holywell"; and the Curtain Theatre would 'appear to have been already in disuse in 1625' (Chappell, 1893, I, p. 181). Fleay says, in disuse in 1623. This gives us the downward, and Golding's Translation the upward, limit. Ben Jonson, in The Devil is an Ass, 1616, mentions a 'Tom o' Bethlem'. But this may be another song. Having said so much about the myth of Actæon, I may here record two more allusions to it in Shakespeare: 'Merry Wives', III, II, 44; 'Titus' II, III, 61—5.

III) FURTHER TRACES OF OVID IN SHAKESPEARE'S WORKS.

Holofernes, in Love's Lab. Lost, IV, II, criticizing Biron's verses, says:

> let me supervise the canzonet. Here are only numbers ratified; but, for the elegancy, facility, and golden cadence of poesy, caret. Ovidius Naso was the man: and why, indeed, Naso but for smelling out the odoriferous flowers of fancy, the jerks of invention?

In 'The Shrew', I, I, 31—3, we read:

> Let's be no stoics nor no stocks, I pray;
> Or so devote to Aristotle's checks
> As Ovid be an outcast quite abjured.

Ovid's biography must have been known, in part at least, to Shakespeare, as appears from As You Like It', III, III, 8—10:

> I am here with thee and thy goats, as the most capricious poet, honest *Ovid, was among the Goths.*

Shakespeare's 'VENUS AND ADONIS' is inspired by Ovid's version of the same story in Metam., X, 519—559, and *circa* 704 ad finem.[1] But Shakespeare seems to be also indebted to two other Ovidian fables.

[1] 'The sweete wittie soule of Ovid lives in mellifluous and hony-tongued 'Shakespeare, witnes his Venus and Adonis, his Lucrece, his sugred Sonnets among 'his private friends, etc.' Meres, Palladis Tamia, 1598.

From the HERMAPHRODITUS AND SALMACIS story[1] (Metam., IV, 285—388) he took the reluctance of Adonis; and from the HUNTING IN CALYDON (Metam., VIII, 270ff), the description of the boar (cf. *ante*).

It is worthy of note that the Venus and Adonis sonnets in 'The Passionate Pilgrim', especially Poem VI., contain a clear recollection of the fable of Hermaphroditus.

I may here remark, without insisting on any connection with Shakespeare's poem, that the story of Venus and Adonis had been treated before by Spenser in his Fairy Queen, III, I, 34—8; and that noteworthy allusions to the fable are to be found in Marlowe's 'Hero and Leander' (1st Sest., vv. 11—14), in Lodge's 'Scillaes Metamorphosis', 1589, (cf. the stanzas quoted in the Shakesp. Soc., Papers, 1847, vol. III, 143), and in Robert Greene's 'Never to Late', 1590 (cf. Malone XX, p. 87). H[enry] C[onstable]'s 'Sheepheard's 'Song of Venus and Adonis' which has much in common with Sh.'s poem appeared in 'England's Helicon' in 1600. No earlier edition of Constable's 'Song' is known. It has the appearance of being a tame copy of Shakespeare's popular poem, the first heir of his invention. From Marlowe's and Greene's allusions to the old myth, we see that Shakespeare was not the first to represent Adonis as cold and disdainful of love.

The PHILOMELA AND TEREUS fable (Met. VI) is frequently referred to. In Titus Andromicus, IV, I, 41, we read:

> *Tit.* Lucius, what book is that she tosseth so?
> *Young Luc.* Grandsire, 'tis *Ovid's Metamorphoses*;
> My mother gave it me.
> *Marc.* For love of her that's gone,
> Perhaps she cull'd it from among the rest.
> *Tit.* Soft! see how busily she turns the leaves!
> What would she find? Lavinia, shall I read?
> This is the tragic tale of Philomel,

[1] The following verses from the story of Salmacis (Met. IV, 320f.), in Golding:
' . . . right happie folke are they
'By whome thou camst into this world; right happie is (I say)
'Thy mother and thy sister too, (if any be:) good hap
'That woman had that was thy nurse, and gave thy mouth hir pap.
'But far above all other, far more blisse than these is she
'Whom thou for thy wife and bedfellow vouchsafest for to bee.'—
ought to be compared with 'The Shrew', IV, v, 39—41:
"Happy the parents of so fair a child;
"Happier the man, whom favourable stars
"Allot thee for his lovely bedfellow!"

And treats of Tereus' treason and his rape;
And rape, I fear, was root of thine annoy.
Marc. See, brother, see; note how she quotes the leaves.
Tit. Lavinia, wert thou thus surprised, sweet girl,
Ravish'd and wrong'd, as Philomela was,
Forced in the ruthless, vast, and gloomy woods?
See, see!
Ay, such a place there is, where we did hunt—
O, had we never, never hunted there!—
Pattern'd by that *the poet* here describes,
By nature made for murders and for rapes.

Compare Act II; scenes 3. and 4.—

Like Philomela, who had informed her sister of Tereus's crime by means of some words woven into a piece of cloth,[1] the dishonoured Lavinia, deprived of her tongue, makes known the truth by way of writing. And as Procne serves up before Tereus the flesh of his child, so Titus revenges himself by making Tamora eat of her sons 'baked 'in a pie'.

The story is distinctly referred to in Titus Andronicus, Act II, III, 43; II, IV, 26, 38—43; IV, I, *ut sup*; V, II, 195—6. Other allusions will be found in Cymb., II, II, 44—6; Mids. N. Dr. II, II, 13, Lucrece, 1079, 1128ff.; Sonnet, 102, 7; Pass. Pilgr., 197.

Lear's words (Lear, II, IV, 280—5):

And let not women's weapons, waterdrops,
Stain my man's cheeks! No, you unnatural hags,
I will have such revenges on you both,
That all the world shall—I will do such things,—
What they are, yet I know not; but they shall be
The terrors of the earth . . .

contain a probable reminiscence, conscious or otherwise, of Ovid's verses 610—618, of which I quote:

. . . Progne, fletum sororis
Corripiens 'non est lacrimis hoc' inquit 'agendum,
Sed ferro . . .
. . . . *magnum quodcumque paravi*
Quod sit, adhuc dubito!

The story of PYRAMUS AND THISBE, the subject of the well-known burlesque interlude in Mids. N. Dr., was no doubt familiar to Shakespeare in the account given by Ovid (Metam. IV). The story is also

[1] Comp. Titus, II, IV, 39—43.

told by Chaucer in his Legend of Good Women. Other versions of the fable need not occupy our attention, as Shakespeare does not appear to be under obligations to them. The myth is also alluded to in Titus Andronicus, II, III, 231:

So pale did shine the moon on Pyramus
When he by night lay bathed in maiden blood,—

in The Merchant of Venice, V, I, 6:

In such a night
Did Thisbe fearfully o'ertrip the dew
And saw the lion's shadow ere himself
And ran dismay'd away,—

and in 'Romeo' II, IV, 45:

Laura to his lady was but a kitchen-wench; . . . Thisbe a grey eye or so, but not to the purpose.

The story of JASON, MEDEA, AND ÆSON. (Metam. VII, 1—293.) Medea's incantation (vv. 197*seq.*) is echoed by Prospero in The Tempest, V, I, 23ff., as I have said above.

The circumstantial account of her magical ceremonies (Met., VII, 262—274)[1] probably supplied suggestions for the witch scene in Macbeth (Act IV, sc. I). Medea, as Ovid records, goes round about the altars with dishevelled hair, brews a magic concoction in a cauldron into which she throws Thessalian roots, seeds, flowers, juices, stones fetched from the farthest East, sea-sand, rime or dew caught from the moon, flesh and wings of the owl, entrails of the wolf, scaly skin of the snake, liver of a live stag, head of an old crow, and many nameless things.

Further allusions to the Medea myth we find in Merch. of Ven., I, I, 172; III, II, 244; V, I, 13—14; 2. Henry VI., Act V, II, 59. (cf. Tristia III, 9.)

[1] It seems that Shakespeare had read this passage in the original. In Macbeth, III, v, 23—25, Hecate says:

"Upon the corner of the moon
"There hangs a vaporous drop profound;
"I'll catch it ere it come to ground" etc.

This may be paralleled by the following verses from Ovid (Met. VII, 268), who says of Medea:

'Addit et *exceptas luna* pernocte *pruinas*'.

Golding translates thus: —

'She put thereto a *deaw* that fell upon a *monday* night'.

The drop from the moon was therefore decidedly not taken from Golding.

THE STORY OF LUCRECE. The problem of the source of Shakespeare's Lucrece has been minutely investigated by Dr. Ewig in Anglia XXII (Shakespeare's Lucrece, eine litterarhistorische Untersuchung). He arrives at the following trustworthy conclusion (p. 32): Shakespeare's poem is based upon Livy's version (Bk. I., ch. 57, 58); probably also the Ovidian (Fasti II, 721 ff.) and perhaps too the Chaucerian versions were made use of. Whether Livy's and Ovid's influence is of a mediate or immediate kind, it is impossible to decide with certainty. But, I think, there ought to be no doubt that Shakespeare had recourse to the Latin writers direct.[1]

Further traces of the legend of Lucrece will be found in The Shrew, II, I, 298; Twelfth Night, II, v, 104; II, v, 116; As You Like It, III, II, 156; Titus, IV, I, 63; Macb., II, I, 55, Cymb., II, II, 12.

Shakespeare's mind was richly furnished with the antique MYTHOLOGY,[2] to which we find innumerable allusions, introduced with perfect ease and naturalness, throughout his works. Much of his knowledge of the ancient fables and legends Shakespeare must have acquired through the medium of Ovid's Metamorphoses, a veritable storehouse of the old mythological lore.

Without aiming at exhaustiveness, we may note the following myths, which Shakespeare would find in Ovid's attractive work:—

Narcissus and Echo: Venus, 161; Lucrece, 285; Antony, II, v, 96; Romeo, II, II, 162.

Phaeton: Gentlemen, III, I, 153; Rich. II., Act. III, III, 178; 3. Henry VI., Act I, IV, 33; Act II, VI, 12; Romeo, III, II, 3; — etc.

[1] Dr. Ewig refers on p. 10 (see also p. 32) to Max Koch for a further proof of Shakespeare's acquaintance with Livy. The reference is worthless, as Koch gives no reasons for his statement.—On the other hand, Dr. Ewig takes no notice of what Dr. Furnivall says in his Introd. to the Leopold Sh., p. CXXVI, 'the story 'of Lucrece is fully told in *Barnabe Googe's* Prouerbes of Lopez de Mendoza, englisht 1575, leaves 58—60, 70 bk., 71 bk., from the Tuscan of M. John Galenis'. —I have looked at a copy in the British Museum, dated 1579, but can find no evidence that Shakespeare made use of it.

Shakespeare's 'Venus and Adonis' and 'Lucrece' are discussed by Wyndham, 'Shakespeare's Poems', 1898; by Sachs, Sh. Jahrb., XXV, 132; and by Baynes *ut sup.*). Dürnhofer has written a weak dissertation on the former poem, Halle, 1890.

[2] Delius contributed a paper on Shakespeare's mythology to the 'Shakesp. Jahrb.' XVIII, 81 ff. But the subject is by no means treated exhaustively. Much more could be made of it.—Thomas Cooper's 'Thesaurus Linguae Romanae & Britannicae' contains an appendix of Proper Names with brief notes. Comp., too, Warton IV, 351—2.

Niobe: Hamlet, I, II, 149; Troilus, V, x, 19.

Dædalus and Icarus: 3. Henry VI., Act V, VI, 18—26; 1. Henry VI., Act IV, VI, 55; VII, 16.

Philemon and Baucis: Much Ado, II, I, 99; As Y. Like It, III, III, 10—11:

Proteus: Gentlemen; 3. Henry VI., Act III, II, 192.

Orpheus: Gentlemen, III, II, 78; Merchant, V, I, 80; Henry VIII., Act III, I, 3; Lucrece, 553; Mids. N. Dr., V, I, 49.

Pygmalion: Meas f. Meas., III, II, 47.

Proserpina: Wint. Tale, IV, IV, 116; Troilus, II, I, 37.

Deucalion: Wint. Tale, IV, IV, 442; Coriolanus, II, I, 102.

Daphne: Mids. N. Dr., II, I, 231; Shrew, Ind., II, 59; Troilus I, I, 101.

Hercules, Nessus, and Lichas: Alls Well, IV, III, 281; Antony, IV, XII, 43; Merchant, II, I, 32; Hamlet, I, IV, 83;—etc.

Laomedon's Daughter: Merchant, III, II, 55.—The bloody Centaurs' feast of 'Titus', V, II, 204, refers to the conflict between the *Lapithæ and Centaurs* at the nuptials of Pirithous and Hippodamia. (Metam. XII, 210ff.)

Several of the names, like Autolycus, Lavinia, Chiron, etc., are derived from Ovid's Metamorphoses.

In conclusion I desire to draw attention to some further parellelisms. Cymb. I, III, 1—22 ought to be compared with Metam. XI, 460ff. (Alcyone taking leave of her husband). Ovid's grandiose description of the tempest (Met. XI, 480ff.) seems to have made an impression on Shakespeare's mind. Compare 'The Tempest', I, I. The following verses in Othello, II, I, 188:

> May the winds blow till they have waken'd death!
> And let the labouring bark climb hills of seas
> Olympus-high and duck againe as low
> As hell's from heaven!

come very close to Ovid's verses, Englished thus by Golding:

> One while as from a mountaines top it [the ship] seemed downe to looke,
> To vallies and the depth of hell, another while beset
> With swelling surges round about which neere above it met
> It looked from the bottome of the whirlpoole up aloft,
> As if it were from hell to heaven.[1]

---

[1] Steevens compares a passage in Sidney, Arcadia, Bk. II, ch. 7, p. 131, verso, of Sommer's edition.

Compare, too, 2. Henry IV., Act III, I, 21:

> . . . winds,
> Who take the ruffian billows by the top,
> Curling their monstrous heads and hanging them
> With deafening clamour in the slippery clouds,
> That, with the hurly, death itself awakes.

With Metam., *ut sup.*, (v. 497):

> The surges mounting up aloft did seeme to mate the skie,
> And with their sprinkling for to wet the clouds that hang on hie.[1]

Regarding Ovid's influence on Shakespeare's style and technique in 'Venus and Adonis' and 'Lucrece', compare Baynes and Wyndham (*ut sup.*).

VIRGIL.

'The Rape of Lucrece' contains evident marks of indebtedness to the Second Book of the *Æneid* for some details in the description of Troy (Lucr., vv. 1366 ff.). Sinon is painted altogether like Virgil's Sinon. Compare, too, Titus Andronicus V, III, 80—7.

From the First Book of the Æneid (v. 11) Shakespeare quotes the following words:—

> Tantaene animis cöelestibus irae?

in 2. Henry VI., Act II, I, 24. In the same play (IV, I, 117) there occurs the following passage

> Gelidus timor occupat artus,—

which the commentators illustrate by Æneid VII, 446:

> Subitus tremor occupat artus.

But this same latter phrase occurs also in in *Ovid's* Metamorphoses III, 40; and *Lucan* I, 246:

> gelidus pavor occupat artus[2]

comes still closer to Shakespeare's words.

For the figure of the Harpy introduced in The Tempest, III, III, Shakespeare may be under obligation to Æneid III. Compare especially the following lines:

---

[1] Arthur Brooke, in his Romeus and Juliet, v. 1361, says:
'As when the winter flawes with dredfull noyse arise,
'And heave the fomy swelling waves up to the starry skies'.
Compare, too, Lucan's 'Nubila tanguntur velis et terra carina' (v. 642).

[2] I have also compared old editions of Lucan's Pharsalia, published in 1569 and 1574.—What can the *Pine* in the Folio of 1623 mean? Did the compositor see and misread *Latine*, or an abbreviation of it, in the MS. from which he printed?

Ariel . . . the elements,
Of whom your swords are temper'd, may as well
Wound the loud winds, or with bemock'd at stabs
Kill the still-closing waters, as diminish
One dowle that's in my plume: my fellow-ministers
Are like invulnerable—

with Aeneid III, 234 ff., translated thus by Phaer (ed 1584):

Their swords by them they laid . . .
And on the filthy birdes they beat, that wild sea rocks do breede,
But fethers none do from them fall, nor wound for stroke doth bleede,
Nor force of weapons hurt them can.

Regarding

TERENCE

see above under Lily's Grammar.

Had Shakespeare read

HORACE?

I have already discussed *Integer vitae*, etc. under the heading of Lily's Grammar. *Ira furor brevis est*, Timon, I, II, 28, is from Epist. I, II, 62. But the phrase seems to have enjoyed proverbial currency. In Love's Labour's Lost, IV, II, 104, Holofernes remarks: "or rather, "as Horace says in his—" Here Holofernes cuts short his own sentence.

PLAUTUS.

The subject of The Comedy of Errors, as is well known, is the same as that of the *Menæchmi* of Plautus. Probably Shakespeare had no need to read the Roman play in the Latin original. Considerable as Plautus's influence was on the Elizabethan drama—a subject which has not been worked out in detail yet—the *Menæchmi*, being peculiarly adapted for representation, must have been produced on the London stage, before Shakespeare tried his hand at the play. This we might be fully warranted in inferring on general *a priori* grounds, even if we did not possess a notice of the performance of 'The Historie of Error' at Hampton Court in 1576—7.[1] The Comedy of Errors is therefore probably a *rifacimento* of an older play.

[1] 'The Historie of Error, showen at Hampton Court on New yeres daie at 'night [1576—7]; enacted by the children of Powles' (Malone, III, p. 387). Ward, in his Hist. of Engl. Dram. Lit. vol. II, p. 74, cautions us not to be too hasty in identifying this play with the 'pre-Errors'. He refers us to Bacon's Advancement

In the 1623 folio we find the names Antiphol $\begin{Bmatrix} i \\ u \end{Bmatrix}$ s $\begin{Bmatrix} \text{Erotes} \\ \text{Errotis} \end{Bmatrix}$ and A. 'Sereptus'. The latter is a corruption of 'Surreptus', the surname of the one Menæchmus of Plautus. 'Erotes' looks like a compromise between '*Erot*ium', the name of the courtezan in Plautus, and 'Sosicl*es*', the surname of the other Menæchmus.

The suggestion that the designations found in the first folio were taken over from the old pre-Shakespearean play seems very plausible.

Whether the translation of the Menæchmi by W. W. (probably William Warner) which appeared in 1595, having been licensed for printing on the 10th of June 1594, was ready about 1591, the probable date of the Comedy of Errors, is uncertain. Malone observes: 'from 'the printer's advertisement to the reader, it appears that, for some 'time before [1595], it had been handed about in MS. among the 'translator's friends'. But there is no internal evidence of Shakespeare having used this translation.

The play mentioned in the 'Gesta Grayorum' as the 'Comedy of 'Errors like to Plautus his Menechmus', and acted in Gray's Inn on Dec. 28., 1594, must have been Shakespeare's drama.

Some *motifs* of Shakespeare's Comedy of Errors are derived from yet another Plautine Comedy: the *Amphitruo*, which supplied hints for the twin servants; as well as for several farcical scenes and situations. Mercury, for example, 'keeps the real Amphitruo out of 'his own house, while Jupiter, the sham Amphitruo, enjoys the real 'one's wife, Alcmena'. This is like Act III, scene 1. of the Comedy of Errors, where Antipholus of Ephesus cannot gain admittance into his house, while his brother and his own wife are at dinner within. The doubts which the Syracusan Dromio is led to entertain regarding his own identity (Act II, II) are from the same play of Plautus.[1]

The names Grumio and Tranio for the servants in The Taming of the Shrew occur in Plautus's *Mostellaria*, where two slaves bear the same names, but there the resemblance ends. The mistaken identities in Twelfth Night are thoroughly Plautine in character and

of Learning and to Burton's Anat. of Mel. for the term 'comedy of errors'. But these two works seem to be too late to prove anything. In 1583, Jan. 6., there was produced before Queen Elizabeth by the Lord Chamberlain's Servants: 'A 'Historie of Ferrar'. Dyce and others propose to read 'Error' for Ferrar. But Fleay queries: 'Ferrara or written by Ferrars?'

[1] Further resemblances are noticed by Paul Wislicenus in 'Die Literatur', 1874, Nr. 1 and 3. Comp. Shakesp. Jahrb. IX, p. 330.

do not appear to be sufficiently accounted for by reference to the source of the plot.

SENECA.

On the influence of Seneca on Shakespeare and the Elizabethan tragedy generally we possess a treatise by Dr. Cunliffe.[1] Unhappily the writer does not pay sufficient attention to indirect influences which may have acted on our poet.[2] Plautus and Seneca were accounted the best for classical Comedy and Tragedy by the Elizabethans. It is evidently significant that Shakespeare should make a reference to the two Roman dramatists in Hamlet, II, II, 419, though through the mouth of Polonius, who is made to say:—

Seneca cannot be too heavy, nor Plautus too light.

We should thus have reason to suppose that Shakespeare had read an author, who, as I have said, was considered to be the model of classical tragedy. This supposition receives confirmation from some traces of Seneca in the works of the great dramatist, especially in Titus Andronicus. The following passage in this play (Titus IV, I, 82):

Magni Dominator poli,
Tam lentus audis scelera? tam lentus vides?

is a quotation, somewhat modified, from *Hippolytus*, l. 671 (Act II):

Magne regnator deum,
Tam lentus audis scelera? tam lentus vides?

The words in Titus Andronicus, II, I, 133—5:

Sit fas aut nefas . . .
Per Styga, per manes vehor,

have been compared with *Hippolytus*, 1180:

Per Styga, per amnes igneos amens sequar![3]

The subject and style of the whole play of Titus Andronicus is thoroughly Senecan in character, though it ought to be remembered that it is not the first of the tragedies of blood and revenge of the Elizabethan period. Prof. Brandl thinks that the horrible feast in Titus Andronicus, V, II and III, contains reminiscences of *Thyestes*.[4]

[1] The Influence of Seneca on Elizabethan Tragedy. An Essay by John W. Cunliffe. London. 1893.

[2] Compare Robertson, Montaigne and Shakspere, pp. 73—81.

[3] Shakesp. Jahrb., IV, 65.

[4] Götting. Gelehrte Anzeig., 1891, Nr. 18, pp. 722—5.

There is a striking, though not conclusive, parallelism between Richard III., Act II, III, 42—4:

By a divine instinct men's minds mistrust
Ensuing dangers; as, by proof, we see
The waters swell before a boisterous storm, —

and *Thyestes* 958—961:

Mittit luctus signa futuri
Mens, ante sui praesaga mali.
Instat nautis fera tempestas,
Cum sine vento tranquilla tument.

Prof. Brandl[1] points out a remarkable resemblance between the first Monologue of Seneca's *Medea* and Macbeth I, v, 41—55. Macbeth's wife, a veritable Medea herself, says:

Come, you spirits
That tend on mortal thoughts, unsex me here,
And fill me from the crown to the toe top-full
Of direst cruelty! make thick my blood;
Stop up the access, and passage to remorse,
That no compunctious visitings of nature
Shake my fell purpose, nor keep peace between
The effect and it! Come to my woman's breasts,
And take my milk for gall, you murdering ministers,
Wherever in your sightless substances
You wait on nature's mischief! Come, thick night,
And pall thee in the dunnest smoke of hell,
That my keen knife see not the wound it makes,
Nor heaven peep through the blanket of the dark,
To cry 'Hold, hold!'

Compare Medea's Monologue, from which I quote the following verses in the English Translation, printed in 1581:

. . . O Juno . . . O Pallas . . . O Titan . . . Hecate . . .
And yee on whom Medea may with safer conscience call,
O Dungeon darke, most dreadfull den of everlasting night,
. . . I conjure you, O grisly Ghostes appeare . . .
If any lusty lyfe as yet within thy soule doe rest,
If ought of auncient corage still doe dwell within my brest,
Exile all foolysh Female feare, and pity from thy mynde,
And as th'untamed Tygers use to rage and rave unkynde,
. . . permit to lodge and rest,
Such salvage brutish tyranny within thy brasen brest.

[1] The Schlegel-Tieck Translation, ed. 1899, vol. VI, 138.

What ever hurly-burly wrought doth Phasis understand,
What mighty monstrous bloudy feate I wrought by Sea or land:
The like in Corynth shalbe seene in most outragious guise. etc.

The Pre-Hamlet was largely influenced by Seneca's dramas, as Nash tells us in his Epistle prefixed to Greene's Menaphon, 1589: 'English *Seneca* read by candle light yeeldes manie good sentences, 'as Bloud is a begger', and so foorth: and if you intreate him faire 'in a frosty morning, he will affoord you whole *Hamlets*, I should say handfulls of tragical speaches.' The Hamlet Tragedy, as we have it, still bears marks of Senecan influence. The appearance of the Ghost crying for Revenge is due to the Roman tragedian. Madness, murder, the guilty wife are all *motifs* which pervade the dramas of the Seneca; but they are also present, or fore-shadowed, in Saxo-Belleforest.

LIVY.

See under Ovid, the legend of Lucrece. (A translation of Livy by A. Nevill was entered on the Stationers' Registers in 1577 (Arber's Transcr. II, 312). Malone, I, 385, mentions a translation by Anth. Cope, 1545; and one by Holland in 1600.)

PLINY.

With Pliny's *Historia Naturalis* Shakespeare seems to have been acquainted in some form or other. Philemon Holland's English translation appeared in 1601. But if the passage in As You Like It (1599) quoted below contain a reminiscence of Pliny, Shakespeare must have familiarized himself with the Natural History before the publication of Holland's translation. This he may have seen in MS. Or did he acquire a knowledge of Pliny through some other medium?

Book I. is nothing but a table of contents of the whole work. Book II. contains an interesting account of the world, the earth, and the sea, and their wonders.

I) The eighty-third chapter, which has the heading 'Monstrous 'Earthquakes seene never but once', records 'a great strange wonder 'of the Earth', which happened 'whiles L. Martius and Sex. Iulius were 'Consuls':

for two hilles encountred together, charging as it were, and with violence assaulting one another, yea and retiring againe with a most mightie noise.

Compare As You Like It, III, II, 194:

> O Lord, Lord! it is a hard matter for friends to meet; but mountains may be removed with earthquakes and so encounter.

II) Book II., chapter 97. concludes with the following sentence:

> And the sea Pontus evermore floweth and runneth out into Propontis, but the sea never retireth backe againe within Pontus.

Compare Othello, III, III, 452:

> *Iago.* Patience, I say; your mind perhaps may change.
> *Oth.* Never, Iago. Like to the Pontic sea,
> Whose icy current and compulsive course
> Ne'er feels retiring ebb, but keeps due on
> To the Propontic and the Hellespont,
> Even so my bloody thoughts, with violent pace,
> Shall ne'er look back, ne'er ebb to humble love,
> Till that a capable and wide revenge
> Swallow them up.

III) Shakespeare is supposed to have derived his knowledge of the Nilometer, to measure the fall and the rise of the waters of the Nile (Antony, II, VII, 20 ff.), from Pliny, Book V, chapt. 9, or from Leo's History of Africa, translated by John Pory, in 1600 (see Malone's Var. Ed.). But this seems uncertain. Compare a later chapter.

IV) There is a curious coincidence between King Lear, IV, VI, 182—4:

> we came crying hither:
> Thou know'st, the first time that we smell the air,
> We *wawl and cry*—

and the following passage in the Proem to the Seventh Book of Holland's Pliny:

> man alone, poor wretch, she [nature] hath laid all naked upon the bare earth, even on his birth-day, to *cry and wraule* presently from the very first houre that he is borne into this world.

LUCAN.

There is some resemblance between Timon's verses on the curse of the gold (Tim. IV, III) and the following passage from Lucan's *Pharsalia*, bk. 1., vv. 161 *seq.*, in *Marlowe's* translation,[1] published in 1600:

> . . . wealth flow'd,
> And then we grew licentious and rude;
> The soldiers' prey and rapine brought in riot;

[1] vol. III., p. 259 of Bullen's editon of Marlowe. Marlowe translated only the 1st book of the *Pharsalia.*

Men took delight in jewels, houses, plate,
And scorn'd old sparing diet, and ware robes
Too light for women . . .
Quarrels were rife; greedy desire, still poor,
Did vild deeds; then 'twas worth the price of blood,
And deem'd renown, to spoil their native town;
Force mastered right, the strongest govern'd all;
Hence came it that th'edicts were over-rul'd,
That laws were broke, tribunes with consuls strove,
Sale made of offices . . .

But Lucan's words are tame compared with those of Timon. *Perhaps* there was a passage to the same effect in the pre-Timon.

JUVENAL.

Warburton supposed that 'the satirical rogue' of Hamlet, II, III, 198, from whom the hero of the play professes to be quoting, is Juvenal, who gives a description of old age in *Sat.* X, 188. This seems very likely.

For the

GESTA ROMANORUM

comp. Giovanni Fiorentino, *post.*

A NOTE.

Lord Say hath *gelded the common-wealth,* and made it an eunuch.
(2. Henry VI., Act IV, II, 174).

This expression occurs in Cicero, De Oratore, bk. III, 41: Nolo dici morte Africani castratam esse rem publicam. Quintilian quotes it, 'Institutio Oratoria', bk. VIII., 6. I find the expression also in Talaeus's Rhetoric, a book used in schools in the sixteenth century.[1] 'Geld' in the sense of diminishing or curtailing is, however, by no means peculiar to Shakespeare. Compare also 1. Henry IV., Act III, I, 110; Love's Lab. Lost, II, I, 149; Rich. II., Act II, I, 237. A writer in Notes and Queries, April 14. 1900., thinks that the following lines in Henry V., Act III, v, 50:

Rush on his host, as doth the melted snow
Upon the valleys, whose low vassal seat
The Alps doth spit and void his rheum upon—

[1] S. Baynes, Fras. Mag., 1880, Jan., p. 90 f.

allude to the conceit of Furius Bibaculus, quoted as far-fetched by Quintilian (*ut sup.*), and jeered at by Horace (Sat. II, 5, 41):—

Jupiter hibernas cana nive conspuit Alpes.

But on looking at the two passages carefully, I find that, though there is a slight resemblance, the difference is far greater. It is absurd to say 'Jupiter spits upon the Alps with white snow', when you mean 'it snows'. But it does not sound so ridiculous to hear the French King comparing himself to the high Alps, and his enemy to the low valleys which the mountain can spit upon in contempt.

CONCLUDING REMARKS.

As the reader will have noticed, I have made no attempt at drawing a hard and fast line between school-classics and Roman authors whom Shakespeare may have perused in later life. The distinction would be practically impossible.

Taking a final review of the matter already dealt with in the present chapter, we may now safely assert, that Shakespeare's knowledge of the Latin language was considerable, and that he must have read some of the more important Latin authors. Besides the reasons already adduced, there are others which confirm this view. Even Ben Jonson allows that he knew 'small Latin',—where the word 'small' should not be underlined. Malone gives us further, indirect, evidence on the point in his Prolegomena (vol. II., 102), where he refers to letters written by one Sturley (High Bailiff of Stratford in 1596) to Richard Quiney of Stratford (High Bailiff in 1601), a friend of Shakespeare's, whose daughter Thomas Quiney, the son of Richard, married in 1616.[1] These letters by Abraham Sturley are interlarded with Latin sentences, and one is entirely in Latin (Malone II, 561), and surely Sturley would not have written what his brother-in-law could not understand. Moreover, Malone draws attention to a Latin letter (probably a school exercise) written by Richard Quiney, the son. The inference by analogy which Malone draws in Shakespeare's case seems perfectly legitimate.

Alexander Schmidt, in his Shakespeare Lexicon (Appendix), gives a long list of Latin words and phrases in Shakespeare's plays. Further evidence of Shakespeare's knowledge of Latin is supplied by English

[1] A letter written by Richard Quiney, the elder, to Shakespeare is still extant.

words like the following:—disjunct, acerb, sequent, exsufflicate, indign, segregated, cadent, intenible, extirpate, pedascule, pudency, pestiferous, antre, admired Miranda, multipotent, etc.

Lastly, Shakespeare has the ancient mythology and history at his fingers' ends, and throughout his plays and poems we find frequent allusions introduced with ease and naturalness.

GREEK LITERATURE.

On Shakespeare's knowledge of Greek there is less to say. Ben Jonson tells us that the poet *had* small Latin and *less Greek*. What Jonson meant by 'less Greek' each reader must interpret for himself. We have no evidence that Shakespeare read any Greek author in the original. Alexander Schmidt gives only two Greek words in his list: *misanthropos* (Timon, IV, III, 53) which Shakespeare could find in a marginal note to Plutarch's Life of Antonius and in the text of the Life of Alcibiades;[1] and the word *threnos*, a superscription in 'The 'Phoenix and the Turtle', while we have 'threne' in the text, v. 49. Words like *cacodemon* (Rich. III., Act I, III, 144), and *anthropophagi* (Othello I, III, 144), may be taken as indicative of the poet's knowledge of a few simple Greek terms, at least.

PLUTARCH.

Plutarch's celebrated biographies were known to Shakespeare in North's English translation, as Farmer satisfactorily proved. Halliwell-Phillipps, in his Outlines II, p. 285, points out that the great dramatist most likely used the edition of 1595, which is much the same as the first edition published in 1579.[2] The third edition appeared in 1603 with fifteen additional lives, and the fourth in 1612.[3]

On the Lives of *Julius Caesar*, *Marcus Brutus*, and *Marcus Antonius*, Shakespeare founded his 'Julius Caesar'. This play, written about 1601, could not be indebted to North's Life of *Augustus*, which

[1] Skeat, Shakespeare's Plutarch, 1892, pp. 216, 296.

[2] The title of the 1595 edition is this: 'The Lives of the Noble Grecians 'and Romanes, compared together by that Grave Learned Philosopher and 'Historiographer, Plutarke of Chaeronea: Translated out of Greeke into French 'by Iames Amiot, . . . and out of French into English, by Thomas North. Imprinted 'at London by Richard Field for Bonham Norton. 1595.'

[3] It has been asserted, but not proved, that a copy of the 1612 edition, now in the Greenock Library, with the initials W. S., was the poet's property.

was first added to the edition of 1603. According to Skeat, Shakespeare also read the Life of *Cicero.* The biographies of *Marcus Antonius* and *Caius Martius Coriolanus* supplied the materials for 'Antony and Cleopatra' and 'Coriolanus' respectively. 'Timon of Athens', too, is indebted to the Life of *Antonius,* as well as to that of *Alcibiades.*[1] (Compare Skeat's Introd. to Shakespeare's Plutarch, pp. XV ff.)

We also find traces of Plutarch outside of the Roman plays and Timon. Some hints of 'A Midsummer Night's Dream' are clearly derived from the Life of *Theseus,* which stands first in North's work (Skeat, p. XIII). As this play seems to have been written in 1594, the poet must have used the edition of 1579; unless we should suppose him to have seen the first sheets of the new edition, which came from the press of Richard Field, the friend of Shakespeare, and printer of his 'Venus and Adonis' and 'Lucrece'. Skeat points out a considerable number of proper names adopted by Shakespeare from Plutarch. The following are further examples of hints and suggestions obtained from the Greek biographer. Macbeth, III, I, 54—7:

> There is none but he
> Whose being I do fear: and, under him,
> My Genius is rebuk'd; as, it is said,
> Mark Antony's was by Caesar—

apparently contains a reminiscence of a passage in Plutarch (Skeat, p. 181) which was certainly in the poet's mind, when he wrote Antony and Cleopatra, Act II, III, 18—22. Again, in Imogen's bedchamber (see Cymbeline, II, IV, 66 ff.) was to be seen represented in tapestry

> Proud Cleopatra, when she met her Roman,
> And Cydnus swell'd above the banks, or for
> The press of boats, or pride.

Compare Skeat, pp. 174—5 and Ant. and Cleop. II, II, 191 ff. The ominous presages of the "mightiest" Caesar's death related by Plutarch (Skeat, p. 97) are referred to by the great poet in Hamlet, I, I, 113 ff., and 'Caesar', Act II, II. "Cato's daughter, Brutus' Portia" is mentioned in the Merchant of Venice, I, I, 166. These, I think, are the chief points deserving of notice.

[1] Though Shakespeare may have known Painter's novel on Timon, Plutarch, and not Painter, is a source of Shakespeare's play, as is evident from a comparison of the two accounts with the drama.

HOMER.

Dr. Small[1] has adduced arguments in favour of the belief that Shakespeare derived some features of the play of 'Troilus and Cressida', which is probably to be assigned to 1601—2, from Chapman's Translation of Homer, seven books (1. 2. 7—11) of the Iliad having appeared in 1598. Small is certainly correct in his assertion that Shakespeare's play contains Homeric features. But he has not proved that the poet used Chapman's Translation. A translation of the first ten Books of the Iliad by Arthur Hall had appeared as early as 1581. And a play of 'Troy' had been acted in London in 1596. See Henslowe's Diary, where mention is also made of a play 'Agamemnon' 1599, which has been said to be identical with 'Troilus and Cressida', 1599, of Henslowe's troupe (or troupes).[2] These plays may have contained Homeric traits. Fleay identifies the Troy play of 1596 with Heywood's First Part of the Iron Age (pr. 1632), which Ward also assigns to an early period of Heywood's career. Certainly, the resemblances to Shakespeare's 'Troilus and Cressida' are very striking, and all the more important Homeric features of this play, pointed out by Small, appear there. That Shakespeare must have been familiar with the classical story of the Trojan war before Chapman's Translation appeared, is clear from Lucrece, vv. 1366 *seq.*

JOSEPHUS.[3]

In King John, II, I, 378 ff. the Bastard says:

> Do like the mutines of Jerusalem,
> Be friends awhile and both conjointly bend
> Your sharpest deeds of malice on this town . . .
> That done, dissever your united strengths,
> And part your mingled colours once again, etc.

'The reference to 'the mutines of Jerusalem' must have been derived, 'directly or indirectly, from Josephus, who in his Jewish War (V. 6, '§ 4) gives an account of the manner in which the leaders of the 'factions in Jerusalem, John of Giscala and Simon bar Gioras, ceased

[1] The Stage-quarrel. Kölbing's Forschungen I, pp. 164 ff.

[2] Or is 'Agamemnon' based on Seneca's play? If so, probably not *only* on Seneca.

[3] Josephus's History of the Jewish War was originally written in Aramaic, but was afterwards translated by its author into Greek, and it is this version alone which we now possess.

'their assaults upon each other to combine in resisting the Roman 'attack. No translation of Josephus into English appears to have 'existed before 1602, but the spurions Hebrew narrative of Josippon, 'or Joseph ben Gorion, had been translated at least in part by Peter 'Morwyng as early as 1558, and several editions were published before 'the end of the 16th century. From this, if from no other source, as 'Malone has shewn, Shakespeare might have derived his knowledge'. (Wright, Clarend. Press ed., p. 104.) *Perhaps* the story of the siege of Jerusalem had been dramatized and acted on a London stage. Henslowe's diary mentions a play 'Jerusalem' acted 1591. Another play in which this subject may have been treated is called 'tittus and Vespacia'. But on this, as on other questions, the information we get from Henslowe's diary is extremely unsatisfactory. Books on the destruction of Jerusalem were printed by Wynkyn de Worde and by Pynson. No doubt ministers of the gospel would refer to those events in their sermons. So Shakespeare may have easily got his knowledge at second hand.

HELIODORUS.

In Twelfth Night, V, I, 120—3, the following words are placed in the mouth of the duke:

Why should I not, had I the heart to do it,
Like to the Egyptian thief at point of death,
Kill what I love?—a savage jealousy
That sometimes savours nobly.

Here we have a plain allusion to the adventures of the Egyptian robber Thyamis, who, brought to bay by his enemies, purposed to kill Chariclea, the object of his affections; but, luckily for her and the readers of the story, stabbed the wrong person. The romance, of which this is an episode, is entitled Αἰθιωπικῶν βίβλια δέκα, composed by Heliodorus, a writer of the third century of the Christian era.

The *Æthiopica* enjoyed a great deal of popularity on the Continent and England alike. An English rendering by Thomas Underdowne from a Latin version was issued in 1569, and again in 1587, 1606, 1622. The romance, of which Theagenes and Chariclea are the heroes, was also dramatized and performed on the English stage as early as 1572—3, as we know from the 'Accounts of the Revels at Court',[1]

[1] See Old Shakespeare Society, 1842, pp. 34—35. Fleay, Hist. of the Stage, p. 20 and note; also Cohn, Sh. in Germany, p. CX.

where we find mention of 'ii speares for the play of Cariclia . . . An 'awltar [altar] for theagines, . . . the picture of Andromadas [for 'Chariclea]'. Stephen Gosson, too, in his 'Plays confuted', *c.* 1582, informs us of the dramatic treatment of the story (Malone III, 40).[1]

MARIANUS.

To von Friesen and Hertzberg[2] we owe the curious discovery of the probable source of Shakespeare's last two sonnets, which show marvellously close resemblance to a Greek epigram of six lines by the Byzantine Marianus, who probably lived in the fifth century. It was Latined in 1529, and several times afterwards. How it reached Shakespeare is a puzzle.

On

ANACREON, LUCIAN, PLATO, etc.,

see later chapters.[3]

[1] The popularity of the Æthiopica on French soil is testified by the ten editions through which Amyot's French translation passed in the second half of the sixteenth century. Racine is said to have known the story by heart, and to have formed the plan of dramatizing it. Nor is there lack of further evidence of its popularity in England. In 1567 James Sandford published a short account of the story 'gathered for the most part out of Heliodorus a Greeke ,Authour' in his 'Amorous Tales', etc. The first few pages of the Æthiopica were metrified by Abraham Fraunce, and published in this form in 1591. Sidney's Arcadia, we know, was influenced by Heliodorus: and William Warner's 'Pan his Syrinx', 1585, was written somewhat in the manner of the Greek romance. I find the story alluded to by Greene (Huth Libr. ed., II, 67, 91; IX, 80), Sidney (Apology, repr. by Arber, p. 28), and Reginald Scot (Disc. of Witchcr. ed. by Dr. Nicholson, p. 503). These allusions, which occur to me, are probably only a few out of many.—Further information on Heliodorus, than I can give here, may be found in Rohde, Der Griechische Roman, 1876; Oeftering, Heliodor und seine Bedeutung für die Literatur, Berlin, 1901; Dunlop, Hist. of Fiction. Underdowne's Translation appeared in the 'Tudor Translations', vol. 5, 1895.

[2] Shakespeare Jahrbuch XIII, 158—162.

[3] In Part II. of his paper (see *ante*, p. 11 note), Mr. Collins will endeavour to show that Shakespeare was acquainted with the Greek classics through Latin translations, which certainly did exist. He has already pointed out a noteworthy parallelism between '*Troilus and Cressida*', III, III, and Plato's *Alcibiades* I. I hope I shall be able to touch upon some of the more important points suggested by Mr. Collins in a later chapter 'On borrowed thoughts in Shakespeare'. Mr. Collins will, I am confident, also discover coincidences between Shakespeare and Greek classics of which Latin translations did *not* exist.

APPENDIX TO CHAPTER 1.

I—. A Reprint of Pages 1. and 2. and part of p. 3. of Lily's Grammar, ed. 1577.

AN INTRODUCTION OF THE EYGHT PARTES OF LATINE SPEACHE.

In Speach be these eyght partes folowing.

Noune, Pronoune, Verbe, Participle,	declined.	Adverbe, Coniunction, Preposition, Interiection,	undeclined.

OF THE NOUNE.

A Noune is the name of a thing, that may be séene felt, hearde, or understande: As the name of my hande in Latine is *Manus*: the name of an house is *Domus*: the name of goodnesse is *Bonitas*.

Of Nounes, some be Substantives, and some be Adiectives.

A Noune Substantive is that standeth by himselfe, and requireth not an other woorde to be ioyned with him: as *Homo*, a man. And it is declined with one Article: as *Hic magister*, a myster. Or else with twoo at the most: as *Hic & hæc parens*, a father or mother.

A Noune Adiective is that can not stande by himselfe, but requireth to be ioyned with an other woorde: as *Bonus*, Good. *Pulcher*, Fayre. And it is declined eyther with thrée Terminations: as *Bonus, bona, bonum:* or else with thrée Articles: as *Hic, hæc, & hoc Fœlix*, Happy. *Hic & hæc levis, & hoc leve*, Light.

A Noune Substantive eyther is proper to the thing that it betokeneth: as *Eduardus*, is my proper name, or else is common to mo: as *Homo*, is a common name to all men.

NUMBERS OF NOUNES.

In Nounes be twoo Numbers, the Singular, and the Plurall. The Singular Number speaketh of one: as *Lapis*, a stone. The plurall number speaketh of mo than one: as *Lapides*, Stones.

CASES OF NOUNES.

Nounes be declined with sixe cases, Singularly, and Plurally, the Nominative, the Genitive, the Dative, the Accusative, the Vocative, and the Ablative.

The Nominative case commeth before the Verbe, and aunswetreh (*sic*) to this question, Who or what: as *Magister docet*, The mayster teacheth.

The Genitive case is knowne by this token Of, and aunswereth to this question, Whose, or where of: as *Doctrina magistri*, The learning of the mayster.

The Dative case is knowne by this token To, and aunswereth to this question, to whome, or to what: as *Do librum magistro.* I give a booke to the mayster.

The Accusative case followeth the Verbe, and aunswereth to this question, whome or what: as *Amo magistrū*, I love the mayster.

The Vocative case is knowne by calling or speaking to: as *O magister*, O mayster.

The Ablative case is commonly ioyned with Prepositions serving to the Ablative case: as *De magistro*, Of the mayster. *Coram magistro*, Before the mayster.

Also In, with, through, for, from, by, or then, after the Comparative degrée, be signes of the Ablative case.

ARTICLES.

Articles are borowed of the Pronoune, and be thus declined.

Singulariter.	Pluraliter.
Nominativo hic, hæc, hoc.	*Nominativo hi, hæ, hæc.*
Genitivo huius.	*Genitivo horum, harum, horū.*
Dativo huic.	*Dativo his.*
Accusativo hunc, hanc, hoc.	*Accusativo hos, has, hæc.*
Vocativo caret.	*Vocativo caret.*
Ablativo hoc, hac, hoc.	*Ablativo his.*

GENDERS OF NOUNES.

Genders of Nounes be seven: the Masculine, the Feminine, the Neuter, the Commune of twoo, the Commune of thrée, the Doubtfull, and the Epicene.

The Masculine Gender is declined with this Article *Hic:* as *Hic vir*, à man.

The Feminine Gender is declined with this Article *Hæc:* as *Hæc mulier*, a woman.

The Neuter gender is declined with this Article *Hoc:* as *Hoc Saxum*, a stone.

The Commune of twoo is declined with *Hic*, and *Hæc:* as *Hic & hæc Parens*, a father or mother.

The Commune of thrée is declined with *Hic, hæc, and Hoc*: as *Hic, hæc & hoc Fœlix*, Happy.

The Doubtfull Gender is declined with *Hic* or *Hæc:* as *Hic vel hæc Dies*, a Day.

The Epicene Gender is declined with one Article, and under that one article both kindes be signified; as *Hic passer*, a Sparrow. *Hæc aquila*, an Aegle, both hée and shée.

THE DECLENSONS (*sic*) OF NOUNES.

There be five Declensons of Nounes. etc. etc.[1]

II—. A Note on the Sententiæ Pueriles.

The *Sententiae Pueriles*, mentioned so frequently as a Shakespearean school-book ever since Malone wrote his observations on the point (Var. Ed. II, 104), deserves a passing notice. It is a little manual consisting of brief Latin sentences collected from divers authors by Leonhard Culmann of Krailsheim and completed probably not long before 1544. The '*sententia*[*e*] '*pueriles* in laten' was entered on the Stationers' Registers in 1569-70 and is referred to in a document of 1584 (Arber I, 418; II, 789). In 1612 the book was translated into English by John Brinsley. Halliwell-Phillipps, in his 'Outlines of the Life of Shakespeare', p. 53, writes: 'The Sententiae 'Pueriles was, in all probability, the little manual by the aid of which he 'first learned to construe Latin, for in one place, at least, he all but literally 'translates a brief passage, and there are in his plays several adaptations 'of its sentiments.' I cannot say exactly what the 'brief passage' is, which Halliwell-Phillipps refers to. The following are some sentences which have a resemblance to passages in Shakespeare. But they are so general in character, that we can scarcely infer anything definite from them.

Belli exitus incertus. Compare Coriol., V, III, 140:
"Thou know'st, great son,
"The end of war's uncertain."

Doloris medicus tempus. Comp. Gent. of Ver., III, II, 15: "A little time, "my lord. will kill that grief." Comp, too, Act III, I, 243; and Cymbeline III, v, 37: "The cure whereof, my lord, 'Tis time "must do."

[1] I have compared a copy of Lily's Grammar published in 1566—7, in the Bodleian Library; but I can find no difference, apart from variations in spelling, between this and the copy of the edition of 1577 in the British Museum. The Latin part is, however, bound before the English portion, which is exceptional. A copy of 1572, of the Bodleian, is also identical with the copy I am quoting from.

Varia et mutabilis semper fœmina. Compare 1. Henry IV., Act II, III, 111: "constant you are, But yet a woman."

Somnus mortis imago. Compare Cymb., II, II, 31: "O sleep, thou ape of "death;" or Macbeth, II, III, 81: "sleep, death's counterfeit."

A curious difficulty in connexion with the *Sententiae Pueriles* is presented by the following passage in George Peele's 'Edward I': "'Tis an 'old said saying, I remember I read it in Cato's *Pueriles,* that *Cantabit 'vacuus coram latrone viator;* a man's purse-penniless may sing before a 'thief.' Now, this quotation occurs neither in Cato's *Disticha de Moribus* nor in the *Sententiae Pueriles.* But it occurs in Lily's Grammar (Part II., Syntax of Adverbs), where Juvenal is mentioned as the author of the phrase. Peele, therefore, seems to have suffered a *lapsus calami.*

III—· A Note on the ABC with the Catechism.

No Elizabethan ABC Book with the Catechism seems to be extant, though thousands, if not millions, were produced. The Bodleian Library possesses a fragment of four leaves printed in 1549. In the library of Saint Cuthbert's College, Ushaw, is an ABC of the year 1553 (?). The next extant copy of an ABC with the Anglican Catechism is, so far as I am aware, of the year 1680, in the Bodleian. The catechism in both copies of the Bodleian Library is the short church catechism, as contained in the Common Prayer Book.[1] Now, Mr. W. H. Allnutt, in his Introduction to an old ABC in Latin (see *post*), makes the following observation: 'c. *1585*? The ABC 'with the Catechisme, printed by the assigns of John Day. A fragment 'of four leaves in the Library of Worcester College, Oxford.' This entry seemed to me so important that I made personal enquiries at Worcester College. Mr. Daniel, however, as well as the Librarian, Mr. Pottinger, informed me, that Mr. Allnutt's statement was, so far as they either o them knew, erroneous. Mr. Allnutt could no longer give information on the point. He had joined the majority just a fortnight before I came to Oxford, and his mouth was closed for ever. Mr. Daniel has been so obliging as to make diligent search, but without success yet.

Meanwhile we must rest satisfied with what information we can derive from those glorious records of the Stationers' Company and other documents, which have escaped the tooth of time:—

Under the date 1561-2 we find the following entry: 'Recevyd of 'Thomas purfoote for his lycense for pryntinge of *an A.bc for chyldren.* 'iiij d.' (Arber, Transcr. I, 182). In 1577 John Day and his son Richard

[1] Nowell's 'Small Catechism' in English is nearly the same as the small Church Catechism, but has some additional matter. However, no copy is known of it earlier than 1574 (?), see Dict. of Nat. Biogr.—Traces of the catechism in Shakespeare's works will be pointed out in a later chapter on the Bible and the Prayer Book.

Day obtained Letters Patent for the sole privilege to print *"the ABC with "the little Catachisme sett forth by her Maiesties Iniunctions for the Instruccon "of Children"*. And in 1582 one Roger Warde is proceeded against in the Star Chamber for an infringement of this Patent, when he confesses to printing 'Twentie Reames of paper' = 10,000 copies. In 1585 another case is brought into the Star Chamber against Thomas Dunne and Robarte Robenson for printing "Tenn Thousand of the bookes cauled *the A.B.C. "with the lyttell Cathechisme* in Englishe". In 1620 the "A.B.C with the "Cathechism" was entered in the Stat. Reg. along with many other school-books. (See Allnutt, *ut infra*, p. 12; Arber, Stat. Reg. I, 111; II, 753, 790; and III, 670.)

For more information, see Mr. H. Bradshaw's paper on 'the ABC as an 'Authorised School-book in the Sixteenth Century', in the Cambridge Antiquarian Communications, vol. III., 1875, pp. 363—373; — E. S. Shuckburgh's Preface to his facsimile reprint of an ABC of *c.* 1538 (London, 1889); — and W. H. A[llnutt], 'An Early Sixteenth-Century ABC in Latin', 1891.[1]

The "ABC with the smaller Catechism", I ought to add, was printed until very recently in Scotland (cf. Tuer, *ut sup.*, p. 373), where the catechism is, of course, that of the Presbyterian Church, beginning with the famous question: "What is the chief end of man?". On this point see A. F. Mitchell, 'Catechisms of the Second Reformation', 1886.

[1] Privately printed, alas! Not even the British Museum has a copy. The Bodleian Library, of which Mr. Allnutt was a member, possesses only the Introduction to the reprint.

CHAPTER 2.

MODERN CONTINENTAL LITERATURE.

FRENCH AUTHORS.

Whether Shakespeare knew French, is the first question we must try to answer. 'King Henry V.', as is well-known, contains numerous French words and phrases and one entire scene in that language. Over and above this, we find a considerable number of French fragments in other plays, which Schmidt has collected in his Shakespeare-Lexicon. His list goes to show that of the modern continental languages French must have been the most familiar to the great dramatist. It is, of course, possible that he received assistance from some one conversant with French, when penning Henry V.; but there is no ground for doubting his acquaintance with the language. It is curious to note, that, while Farmer attributed the French dialogue of the play to another hand, other critics made Shakespeare a bad pronouncer of the language[1] on the score of the following passage, Act IV., iv, 17—21:

French Soldier: Est-il impossible d'échapper la force de ton *bras?*
Pistol: Brass, cur!
Thou damned and luxurious mountain goat,
Offer'st me *brass?*

Impossible though this pun may be with our modern pronunciation, in Shakespeare's days there could be no objection to it, as the final *s* was still audible in words before a pause, though gradually dropping out of use. See Thurot, 'De la Prononciation Française depuis le Commencement du XVIe siècle' (1881—3, II, 35—36), a work to which I was referred by Professor Dr. Tobler, of the University of Berlin,

[1] Even Farmer is so inconsistent as to support this view.

one of the first authorities, if not the first, on Romance Philology. This learned professor, having, very obligingly, also examined Act III, sc. IV, and Act V, II, of Henry V. for me, with regard to their correctness, observes: 'If in Act III, IV, *foot* and *gown* sound nearly like '*foute* and *con*, we ought to bear in mind that the lady-in-waiting 'is a French woman, who may pronounce the English in a ludicrous 'manner. In Act V, II, the French of the king is clumsy, but this is 'evidently the poet's intention'.[1]

Knowing, therefore, that Shakespeare's text contains no blunders, we may rest in measureless content. That the great poet, who commanded his own language, as no one before or after him, could have found no difficulty in acquiring a foreign language like the French, cannot be doubted. But whether he devoted much time and energy to a deeper study of it, is another question, which I am inclined to answer in the negative, seeing that no markedly great influence of French literature is discernible in his writings.

MONTAIGNE.

Michel de Montaigne's Essays Shakespeare must have had lying before him on his table, when he penned 'Tempest', Act II, I, 143 ff., where Gonzalo repeats, with a fine touch of humour, the ideas of the French author on a *natural* commonwealth. In the 'Archiv für das Studium der neueren Sprachen', vol. CVII., p. 181, I have endeavoured to make good the belief (held and expressed long before) that Shakespeare used Florio's translation of the *Essais*, in the thirtieth chapter of which, entitled 'Of the Caniballes', we find the following passage, paraphrased by the poet:—

> *All things* (saith *Plato*) *are produced either by nature, by fortune, or by arte. The greatest and fairest by one or other of the two first, the least and imperfect by the last.* Those nations seeme therefore so barbarous unto mee, because they have received very-little fashion from humane wit, and are yet neere their originall naturalitie. The lawes of nature do yet commaund them, which are but little bastardized by ours. And that with such puritie, as I am sometimes grieved the knowledge of

[1] The rhyme 'pardonne moi': 'destroy' (Richard II., Act V, III, 119—120) and the quibble with 'no point' in Love's Labour's Lost, II, 190 and Act V, II, 277, may be defended on the ground, that phrases and words like these in question had become fashionable and lost their native sound. Compare Romeo, II, IV, 35: "these fashion-mongers, these perdona-mi's" (Folio: 'pardon-mee's'). Cf. Hunter, New Illustrations, II, 321 *seq.*

it came no sooner to light, at what time ther were men, that better than we could have judged of-it. I am sorie, *Licurgus* and *Plato* had it not: for me seemeth that what in those nations wee see by experience, doth not onelie exceede all the pictures wherewith licentious Poesie hath prowdly imbellished the golden age, & al hir quaint inventions to faine a happy condition of man, but also the conception & desire of Philosophie. They could not imagine a genuitie so pure and simple, as we see it by experience; nor ever beleeve our societie might be maintained with so little arte and humane combination. It is a nation, would I answere *Plato*, that hath no kinde of traffike, no knowledge of Letters, no intelligence of numbers, no name of magistrate, nor of politike superioritie; no use of service, of riches, or of poverty; no contracts, no successions, no dividences, no occupation but idle[1]; no respect of kinred, but common, no apparrell but naturall, no manuring[2] of lands, no use of wine, corne, or mettle. The very words that import lying, falshood, treason, dissimulation, covetousnes, envie, detraction, and pardon, were never heard of amongst-them. How dissonant would hee finde his imaginary common-wealth from this perfection? (Florio, p. 102.)[3]

For convenience' sake I place here the whole passage from The Tempest, II, I, 143 *seq.*—

Gonzalo. Had I plantation of this isle, my lord,—
Antonio. He'ld sow't with nettle-seed.
Sebastian. Or docks, or mallows.
Gon. And were the king on't, what would I do?
Seb. 'Scape being drunk for want of wine.
Gon. I' the *commonwealth* I would by contraries
Execute all things; for *no kind of traffic*
Would I admit; *no name of magistrate;*
Letters should not be known; riches, poverty,
And use of service, none; contract, succession,
Bourn, bound of land, tilth, vineyard, none;

---

[1] Shakespeares 'line, "No occupation: all men idle, all", is unquestionably 'a version of Florio, whose "no occupation but idle" did not seem to mean what 'it stood for—"nulles occupations qu'oisifves", no occupations but idle ones'. The Tempest, edited by H. Morley in Cassel's National Library, p. 164.

[2] manure = till.

[3] I am quoting from Florio's first edition, 1603. The italics have been retained, and more important words and phrases made prominent by means of spacing out of the letters. The edition of 1632 (reprinted in the Tudor Translations, 1892—3) differs slightly from that of 1603—'dividences' is changed to 'partitions', instead of 'dissimulation' we have 'dissimulations'.

No use of metal, corn, or wine, or oil;
No occupation; all men idle, all;
And women too, but innocent and *pure;*
No sovereignty; —
Seb. Yet he would be king on't.
Ant. The latter end of his commonwealth forgets the beginning.
Gon. *All things* in *common nature* should *produce*
Without sweat or endeavour: *treason,* felony,
Sword, pike, knife, gun, or need of any engine,
Would I not have; but nature should bring forth,
Of its own kind, all foison, all abundance,
To feed my innocent people.
Seb. No marrying 'mong his subjects?
Ant. None, man; all idle; whores and knaves.
Gon. I would with such perfection govern, sir,
To *excel the golden age.*
Seb. God save his majesty!
Ant. Long live Gonzalo!
Gon. And, — do you mark me, sir?
Alon. Prithee, no more: thou dost talk nothing to me.
Gon. I do well believe your highness; and did it to minister occasion to these gentlemen, etc.

In illustration of vv. 159—164, 'All things . . . nature should produce, etc.', I ought to quote the following passages from Montaigne's chapter, the whole of which Shakespeare must have read:—

They yet enjoy that naturall ubertie and fruitfulnesse, which without labouring-toyle, doth in such plenteous aboundance furnish them with all necessary things, that they neede not enlarge their limites. (Florio, p. 104.)

Again (p. 104.):—

They are even savage, as we call those fruites wilde, which nature of hir selfe, and of hir ordinarie progresse hath produced.

Once we know with certainty that Shakespeare read Montaigne, we may readily infer that this original and stimulating author, one of the most advanced of Renaissance thinkers, could not but have exercised further influence on the great dramatist. We possess several books and articles dealing with this problem. Jacob Feis is the author of Shakespeare and Montaigne', 1884,—an extravagant book—, and John Mackinnon Robertson[1] wrote a verbose work with nearly the same title, London, 1897.

[1] A writer perhaps better known now as the author of 'Wrecking the Empire', 1901.—Robertson's 'Montaigne and Shakspere' was reviewed by Prof. Brandl

Both Feis and Robertson agree in attributing to Montaigne a considerable amount of influence on the English poet, and, especially, in finding many suggestions for the character and play of Hamlet in the *Essais*. Though their position appears exaggerated sometimes, we find no difficulty in conceding a general and broad resemblance between the play just referred to and Montaigne's work, both of which are characterized by a reflective, introspective, and occasionally pessimistic tone, or (to use Shakespeare's language) by a pale cast of thought.[1]

The following are some parallelisms, which have been pointed out. But they are hardly conclusive.

> If it [death] be a consummation [=anéantissement] of ones being, it is also an amendement and entrance into a long and quiet night. Wee finde nothing so sweete in life, as a quiet rest and gentle sleepe, and without dreames. (Florio, Bk. III, ch. 12, p. 627.)

Here we have the same idea which we find in Hamlet's famous monologue, 'To be or not to be'. But Montaigne is confessedly repeating Socrates's thoughts on Death, which may have reached Shakespeare by another channel. (Cf. a later chapter on Borrowed Ideas.)

The following passage, too, is illustrative of Hamlet's soliloquy:—

> My consultation doth somewhat roughlie hew the matter, and by it's first show, lightly consider the same: the maine and chiefe point of the worke, I am wonte to resigne to heaven. (Florio, Bk. III, ch. 8, p. 559.)

For further parallelisms, compare Robertson's 'Montaigne and Shakspere'.

Regarding Montaigne's influence on Hamlet, we have to face another problem. In what version would Shakespeare have read the *Essais?* Florio's translation appeared in 1603; while Hamlet was written sometime before, probably in 1600—1. Of course, Shakespeare

in Shakespeare Jahrbuch, 1899.—Feis's work was discussed, rather uncritically, by Karl Blind, in Das Magazin für die Litteratur des In- und Auslandes, Nov., 1884 (also in 'Shakesperiana', Philadelphia, vol. II., p. 469). Sir William Bailey is the author of an unimportant pamphlet on Shakespeare and Montaigne, published for private circulation in 1895, now out of print. Stedefeld's 'Hamlet, ein Tendenzdrama Sheakspeares (sic) gegen die skeptische und kosmopolitische Weltanschauung des M. de Montaigne', 1871, is scarcely deserving of mention. What can you expect of a man, though a Kreisgerichtsrat (as Stedefeld was), who cannot spell the poet's name?

[1] This would also apply in some measure to Lyly's Euphues.

may have used the original French text. Or are we to suppose that Florio's translation was accessible to him in manuscript, before it was printed? We know, for example, that Sir William Cornwallis, an imitator of the French Essayist, saw Florio's MSS., for in his Essays published in 1600—1, Cornwallis writes:

> For profitable Recreation, that Noble French Knight, the Lord *de Montaigne* is most excellent, whom though I have not bene so much beholding to the French as to see in his Originall, yet divers of his peeces I have seene translated: they that understand both languages say very well done, and I am able to sày (if you will take the word of Ignorance) translated into a stile, admitting as fewe Idle words as our language wil endure: It is well fitted in this newe garment, and *Montaigne* speaks now good English: It is done by a fellowe lesse beholding to nature for his fortune then witte, yet lesser for his face then fortune; the truth is, hee lookes more like a good-fellowe, then a wise-man, and yet hee is wise, beyond either his fortune, or education: but his Authour speakes nobly, honestly, and wisely, with little method, but with much iudgement, etc. (Essay 12.)

This is unquestionably a reference to Florio's version. As Florio stood in good connexions with some of the leading *literati* of his time[1], and was a *protégé* of the Earl of Southampton, we might suppose that Shakespeare, too, would have enjoyed Florio's personal acquaintanceship and thus perhaps had access to his MSS.

In the Stationers' Registers we find the following entry under the date 20. October, 1595:

> Edward Aggas. Entred for his Copie under his handes of the Ward[e]nes 'The Essais' of Michaell Lord of Mountene.

This no doubt refers to Florio's translation, first printed in 1603, having been previously licensed, for a second time, in 1600. Florio thus seems to have been busy at his work for a period of eight years.

RABELAIS.

Of the remaining French writers of the sixteenth century the greatest is François Rabelais (d. 1553), the author of the famous and immensely popular romance of *Gargantua and Pantagruel*, between

[1] Daniel, wrote "Prefatory Verses to my deere friend M. John Florio", printed before the Translation of 1603. In the edition of 1611 Daniel addresses him as "my deare brother and friend". Ben Jonson presented Florio with a copy of his 'Volpone' with his own signature. This book is now in the British Museum. But Shakespeare's 'autograph' in a copy of Florio's translation of 1603 (which I saw myself) is a forgery.

which and Shakespeare's dramas it is pleasant to find a link of connexion. Of this work an English translation or adaptation must have existed in the days of Queen Elizabeth, for Joseph Hall, in his 'Virgidemiarum, Sixe Bookes', 1597, Bk. II, Sat. 1, says:—

> But who coniur'd this bawdie *Poggies*[1] ghost,
> From out the *stewes* of his lewde home-bred coast:
> Or wicked *Rablais* dronken revellings,
> To grace the mis-rule of our Tavernings?

'The historie of Gargantua' was licenced by the Stationers' Company on the 4th of December 1594, 'Provided that if this Copie 'doo belonge to anie other, Then this Entrance to be voide'[2]. But an English book on Gargantua must have been current long before this. "E. D.", in his 'Brief and Necessary Instruction', 1572, decries 'the witles devices of Gargantua' with other books of his time. 'Gargantua' is mentioned by Robert Laneham in 1575 as belonging to Captain Cox's library. As all the rest of his books are English, we should expect this one to be so too. In 1577 Hanmer enumerates 'the monstrous fables of Garagantua' in a list of popular English books. However, no shred of an Elizabethan English work on Gargantua has been preserved to us.

The following are traces, or supposed traces, of Rabelais in Shakespeare's works:—

1) The plainest and most direct allusion to the giant hero of the humorous romance is to be found in As You Like It, III, II, 238. Rosalind putting a long list of questions, urges Celia to answer her all these in one word; to which Celia:

> You must borrow me *Gargantua's* mouth first: 'tis a word too great for any mouth of this age's size.

2) It can scarcely be looked upon as accidental that the pedant of Love's Labour's Lost bears the same name as his intellectual

[1] Alluding to a translation of the indecent *Facetiæ* written by Poggio (d. 1459).

[2]) Gargantua his prophesie (=Rabelais's 'Pantagrueline Prognostication'?) was entered on the Stationers' Registers in 1592, on the 6th of April. Another entry (afterwards cancelled), dated 16th of June, no doubt refers to the same work. For Elizabethan allusions to Rabelais's romance, see Smith's translation of Rabelais, 1893, p. XIII; and Rabelais in the 'Tudor Translations', 1900, p. LXXVII. See also Halliwell's folio edition of Shakespeare, vol. VI, p. 191.—I would scarcely recommend a perusal of an article in Shakespeare Jahrbuch, IX, 195 ff.

cousin-german, Thubal Holofernes, the instructor of Gargantua. (See Bk. I, ch. 14.)[1]

3) Edgar, in King Lear, III, VI, 7, says:—

> Frateretto calls me; and tells me *Nero is an angler in the lake of darkness.*

Though neither Rabelais, nor any other author, so far as we know, ever represented Nero as an angler in hell, Trajan is introduced as a fisher of frogs in Hades by the French humourist, while Nero is made a fiddler, Aeneas, *e. g.*, a miller, Cleopatra a hawker of onions, and so forth (Bk. II, ch. 30.).

4) An expression in Othello, I, I, 116—7,

> your daughter and the Moor are now making the beast with two backs—

has been traced to Rabelais, Bk. I, ch. 3, where we read of Gargantua's father:—

> En son aage virile espousa Gargamelle fille du Roy des Parpaill*ons*, belle gouge, & de bonne trongne. Et *faisoient* eux deux souvent ensemble *la beste à deux dos.*[2]

Possibly the phrase was more or less proverbial.

5) Mr. W. F. Smith, the translator of Rabelais, 1893 (p. XIII), connects the following gibberish of Sir Andrew, in Twelfth Night, II, III, 23—25:

> when thou spokest of Pigrogromitus, of the Vapians passing the equinoctial of Queubus—

with the unintelligible speech of Kissbreech before Pantagruel, Bk. II, ch. 11, *ad init:*—

> But to the purpose, there passed between the two Tropics six white Pieces towards the Zenith and a Halfpenny, forasmuch as the Rhiphaean Mountains had this year had a great Sterility of Happelourdes, etc., etc.[3]

Having thus found traces of the greatest of French humourists in our poet's works, we are tempted to believe that a grain of the Rabelaisian Pantagruelism went to the making of some of Shakespeare's comical characters.

[1] B. E. Smith, in his Dictionary of Names, asserts that Holofernes is the name of a conventional character in the Italian comedy, a statement for which I can find no foundation.

[2] Quoted from Les Oeuvres de Rabelais, Lyon, 1584.—Compare also Bk. V, ch. 30: "I saw some Beasts with two Backs". Smith's Transl., vol. II, p. 391.

[3] *ibid.*, I., p. 262.

RONSARD.

Pierre de Ronsard (1524—1585), 'Prince of Poets', as his own generation in France called him, the founder of the classicistic school of French poets, enjoyed considerable estimation among English poets and sonnetists. Queen Elizabeth, evidently charmed with his poetry, 'gave him a diamond, comparing its water to the purity of his verse'. Sidney, Daniel, Drayton, Watson, Lodge, Shakespeare, and others discover traces of Ronsard's influence.[1] 'Most of Ronsard's nine hundred 'sonnets and many of his numerous odes were accessible to Shakespeare 'in English adaptations, but there are a few signs that Shakespeare 'had recourse to Ronsard direct'.[2]

Apart from Shakespeare's Sonnets, the following passages seem to contain echoes of Ronsard:

1) Timon, Act IV, III, 438—445:

I'll example you with thievery:
The sun's a thief, and with his great attraction
Robs the vast sea: the moon's an arrant thief,
And her pale fire she snatches from the sun:
The sea's a thief, whose liquid surge resolves
The moon into salt tears: the earth's a thief,
That feeds and breeds by a composture stolen
From general excrement: each thing's a thief.

Compare Ronsard, Odes, Livre IV, 31:[3]

La terre les eaux va boivant,
L'arbre la boit par sa racine,
La mer salée boit le vent,
Et le Soleil boit la marine,

Le Soleil est beu de la Lune:
Tout boit, soit en haut ou en bas:
Suivant ceste reigle commune
Pourquoy donc ne boirons-nous pas?

---

[1] See G. Wyndham, the Poems of Shakespeare, p. 211; Sidney Lee, Life of Shakespeare, index.

[2] S. Lee, *ut sup.*, p. 111.—John Southern was noted for his unblushing plagiarisms from Ronsard. (See Dict. of Nat. Biogr., s. v. Southern, and Puttenham's Art of English Poesie, Arber's Reprint, pp. 259—260.) But Shakespeare does not seem to be indebted to Southern.

[3] I quote from Ronsard, Œuvres, 1584.

This is a free rendering of Ode 18 (or 19) of Anacreon.

2) This Ode is immediately preceded by another one (ode XXIX.)[1], which begins thus:

> Les Muses *lierent* un iour
> *De chaisnes de roses*, Amour,
> Et pour le garder le donnerent
> Aux Graces & à la Beauté,
> Qui voyans sa desloyauté
> Sus Parnasse l'*emprisonnerent*.

Compare 'Venus and Adonis', l. 110, where Venus boasts of having overswayed Mars,

> Leading him prisoner in a red-rose chain.

ITALIAN LITERATURE.

We are neither in position to assert positively, nor able to deny with certainty, that Shakespeare was master of the Italian language, though, to judge from Schmidt's list in the Appendix to his Shakespeare Lexicon, he cannot have been altogether ignorant of it. But whatever belief we may hold on this point, it is in the Italian literature that we discover, what must be the original, if they are not the immediate, sources of some Shakespearean plots and motives.[2] I do not desire to bind the reader to either view. For my part, I am much inclined to believe that the poet's indebtedness to the Italian authors is of an indirect kind. The plays founded on Italian tales show a more marked divergence from their originals than most of the plays based on known English sources, whether prose or verse; the reason being, I think, that Shakespeare let others do the quarrying and even the rough-hewing of the blocks of Italian marble, to which he gave their final perfect shape. In dealing with his relation to Italian literature, we should keep in mind what Stephen Gosson says (though he may be exaggerating somewhat) in his 'Playes Confuted', 1582:—

> I may boldely say it because I have seene it, that the Palace of pleasure, the Golden Asse, the Æthiopian historie, Amadis of Fraunce, the Rounde

[1] This too is a paraphrase of an Anacreontic ode (no. 30 in some editions).

[2] In The Gentleman's Magazine, Dec. 1895, there is a paper by Mr. C. F. Walters on 'Italian Influence on Shakespeare'.

table, baudie Comedies in Latine, French, Italian, and Spanish, have beene *throughly ransackt to furnish the Playe houses in London.*[1]

BOCCACCIO.

1. The love-story (Posthumus-Imogen) of 'Cymbeline' is founded on Boccaccio, *Decameron*, II., 9.[2] Ohle, in his elaborate treatise 'Shakespeare's Cymbeline', gives reasons for supposing that yet another version of the story was known to the author of the play[3].

ANOTHER SOURCE OF CYMBELINE?[4]

Ohle does not pretend to have discovered this other source; but, he says, it must have been a version, probably English, running nearly parallel with the dramatic treatment of the story in the French *Miracle de Oton, Roy d'Espagne* (*circa* 1380). The chief coincidences, pointed out by Ohle between the two plays, and not occurring in other versions, are these.

First, the *deus ex machinâ* scene in Cymbeline (V, IV) reminds one of the following scene in the *Miracle*, where God appears to Ostes (Otto), the hero of the play, and reveals to him the innocence of his wife:

Ostes prays and repents having turned an infidel (v. 1501 ff.).

Dieu. Mére, et vous, Jehan, alons ment La jus a ce pecheur Oston . . . . . Sa dévote contriccion, Qui de lermes moulle sa face, Me contraint que grace li face. Or sus, trestouz!	*God.* Mother, and thou, John, go we down there to this sinner, Ostes . . . . . His devout contrition, which bathes his face with tears, constrains me to grant him pardon. Come, now, all of you!
Nostre Dame. Mon Dieu, mon pére et mon filz doulz, Nous ferons vostre voulenté. Sus, anges! soiez apresté De tost descendre	*Our Lady.* My God, my father and my sweet son, we shall do thy will.— Come, angels, be prepared to descend anon.

[1] Copied from Hazlitt, the Roxburghe Library, English Drama and Stage, 1869, p 188—9.

[2] Boccaccio's Decameron was translated into French by Laurens du Premierfait 1485, etc., and by Ant. le Maçon, 1545, etc. The first English translation appeared in 1620.

[3] Comp. also Brandl, Schlegel-Tieck Translation, v. X., pp. 4—5.

[4] Comp. wat I have said above in the Preface.

Gabriel. Dame, qui peustes comprendre
Ce que ne peuent pas les cieulx,
Chascun de nous est ententiex
De voz grez faire.

. . . .

Dieu. Ostes, pour la contriccion
Vraie que je voy estre en toy
As recouvré grace. Taiz toy.
A Romme tout droit t'en iras;
La ton pechié confesseras:
Puis qu'a repentence es venuz,
Il le fault, tu y es tenuz,
Ou ce que tu fais rien ne vault.
Oultre, tu as un grant deffault,
Qu'a tort as ta femme hay
Et jusques a mort envay:
Et pour ç'aussi tu la querras,
Et pardon li en requerras.
Plus ne demeure en ceste terre,
Mais a Romme t'en vas bonne erre,
Et fay ce que t'ay divisé.
Je l'ay assez bien avisé.
Sus! alons ment.

Nostre Dame. Avant, anges, etc.

. . . .

Ostes. Pére de consolacion,
Piteux, doulx et misericors,
Ha! sire, quant je me recors
Que des cieulx vous estes oultré
Et a moy vous estes monstré,
Et vostre doulce mére aussi,
Et que je vous ay veu cy,
Bien doy bouche, mains et cuer tendre
A vous loer et graces rendre.
Cy endroit plus ne demourray;
Mais a Romme seul m'en iray
Tout maintenant.

Gabriel. Lady, who couldst comprehend, that which the heavens cannot, each one of us is intent on doing thy will.

[They descend.]

God. Ostes, owing to thy true contrition, which I see in thee, thou hast recovered grace. Hush! Thou shalt go hence straight to Rome; there thou shalt confess thy sin: seeing that thou hast come to repentance, it is necessary that thou shouldst be ruled by it, otherwise thy doings are worthless. Besides, thou art guilty of a great misdeed in wrongly hating thy wife and persecuting her to death: wherefore thou shalt seek her out, and ask her pardon. Stay no longer in this country, but go with all speed to Rome, and do what I have indicated to thee.—I have advised him sufficiently well. Come, go we hence.

Our Lady. On, you angels, etc.

Ostes. Father of consolation, piteous, benign, and merciful. Oh, Lord, when I recall to mind that thou hast descended from the heavens and hast appeared to me and thy sweet mother also, and that I have seen thee here, my mouth, my hands, and my heart must render praise and thanks unto thee. I shall no longer remain here; but shall betake myself to Rome alone immediately.

The reader will notice the resemblance but also the great difference between the above scene and that of Cymbeline. It is possible that the correspondence on which Ohle lays so much stress is accidental, and that the scene in Cymbeline was suggested by the courtmasques, so popular in King James's reign, of which the scenery and the machinery business formed a great attraction.

Another noteworthy coincidence is that the scene of the wager is laid in Rome in both plays. Perhaps too much importance has been attached to this. We ought to bear in mind that, once the story is made to take place in the time of King Cymbeline, a contemporary of Caesar Augustus, the change of scene from Boccaccio's Paris to the seven hilled city seems most natural. Berengier's insinuation (he is the Iachimo of the *Miracle*) to the lady about Ostes's infidelity does not seem to me very noteworthy. In Boccaccio, however, the villain (Ambrogiuolo) does not communicate with the lady at all.

Ohle points out, too, that "in both plays the several characters "and the hero and heroine are brought together again by agency of "a war". This statement ought to be more explicit, because, as it stands, it is likely to be misleading. In the Miracle the king of Grenada along with five other kings is about to march against the Emperor of Rome. The disguised lady, the niece of the king of Grenada, is sent before as a messenger to Rome, and there challenges Berengier. Ostes has arrived on the scene, and overcomes the latter in single combat. Berengier confesses his crime, the messenger discloses her sex, and the war is at an end before it is waged.

Further, it has been pointed out that Berengier boasts of being able to accomplish his object after *two* interviews:

Et vous dy bien que je me vant Que je ne sçay femme vivant Mais que deux foiz a li parlasse Que la tierce avoir n'en cuidasse Tout mon delit. (v. 653.)	And I tell you clearly, that I vaunt myself of not knowing a woman alive, but, if I had spoken to her twice, I would hope to enjoy her at the third time.

Compare Cymbeline, I, IV, 138 ff.:

> I will lay you ten thousand ducats to your ring, that, commend me to the court where your lady is, with no more advantage than the opportunity of a *second conference*, and I will bring from thence that honour of hers which you imagine so reserved.

I would not emphasize this coincidence.

Mr. Herford, in his Introduction to Cymbeline[1], assuming an English version, writes,

---

[1] Eversley Edition, vol. IV, p. 116.

. . . Perhaps, too, the English tradition may have agreed with the German, *Volksbuch*[1] in making the wager originate in a company of 'four merchants' corresponding to Pisanio's four guests of various nationalities.

This conjecture of Herford's is nicely verified by the tale of Frederyke of Jennen, a fragment of which is reprinted by Dr. Furnivall in his edition of Laneham's Letter. A copy of the whole tract (of *c.* 1560) is in the British Museum[2]. From this I quote:

In the yere of our lorde god. M. CCCC. XXIII. It happened that foure ryche marchauntes departed out of divers countreis for to do their marchaundise . . . . . thei were al foure goyng towarde Paris in Fraunce and for company sake they rode al. IIII. into one ynne . . . . The firste was called Courant of Spayne, the secōd was called Borcharde of Fraunce, ye thirde was called John of Forence *(sic)*, & the. IIII. was called Ambrose of Jennen.[3]

Ohle has no hesitation in saying, that Iachimo's comparison of the mole on Imogen's breast to "crimson drops I' the bottom of a "cowslip" was suggested by the *Roman de la Violette* (*c.* 1230), where we read (v. 660):

Et voit Desor sa destre mamelete Indoier cele violete.	And he sees on her right breast the appearance of this violet.[4]

And again (v. 956):

Desour sa destre mamelete A une biele violete.	On her right breast there is a beautiful violet.

After having pointed out the principal reasons which have been adduced in favour of a hypothetical second source, I refer the reader for a further discussion and for information on the various versions of our story and their relation to each other to Ohle, whose work is, however, vitiated by unnecessary complicacy and, occasionally, by an incomplete knowledge of facts. For my part, I am not easily satisfied by anything short of a tangible proof. Ohle's supposition that the names Posthumus, Imogen, and Cloten point to an old English version of our story seems to me scarcely worth discussing. Shakespeare, we

[1] Ein liepliche History und Warheit von Vier Kaufmennern, 4. (Herf.)

[2] Compare Hazlitt, Handbook, 212; Collect. and Notes I, 172; III, 88. The Press mark of the copy in the Brit. Museum is C. 20. c. 42. (6). Of the above mentioned coincidences only the last is to be found in Fred. of Jennen.

[3] As a matter of fact there are more than four persons in Philario's house, in Cymb., 1, IV. But the fact that they are of different nationalities is noteworthy.

[4] Indoier=paraître violet (Godefroy). In Guillaume de Dole (*c.* 1200) the maid has a rose on her thigh (see Mod. Lang. Assoc., Trans. vol. II, 1886, p. 118).

know, altered the names of his sources according to his fancy. Compare, for instance, the names in As You Like It, Winter's Tale, Measure for Measure, Taming of the Shrew, etc. In Cymbeline the above names, together with others, are taken from Holinshed, who supplied the entire historical framework of the play (see Boswell-Stone, 'Shakspere's Holinshed').

As for 'The Tale told by the Fishwife of Standon-the-Green', contained in *Westward for Smelts*, 1620[1], Ohle has not satisfactorily made out its relation to the other versions. The concealment of the contriver of harms in the lady's chamber, whose maid is near at hand, looks like a reminiscence of Boccaccio. In both versions we are in bourgeois society. The absence of the mole on the lady's breast is probably due to the prudery of the narrator[2]. On the other hand the version in Westward for Smelts agrees with Cymbeline, while disagreeing with Boccaccio, in the circumstance that the husband and the wife are re-united by the agency of war, which takes place in England. Just as Iachimo has a letter of introduction by Posthumus, so the villain of our prose tale professes to have recommendations to the lady from her husband. Probably the narrator of the Fishwife's Tale knew both Boccaccio's and Shakespeare's versions.[3]

In conclusion, I draw up the following trial table exhibiting Shakespeare's relation to his sources. I hypothetize an earlier English play, which the great poet revised, and make allowance for a supposed second version, which I call x, whatever it was.

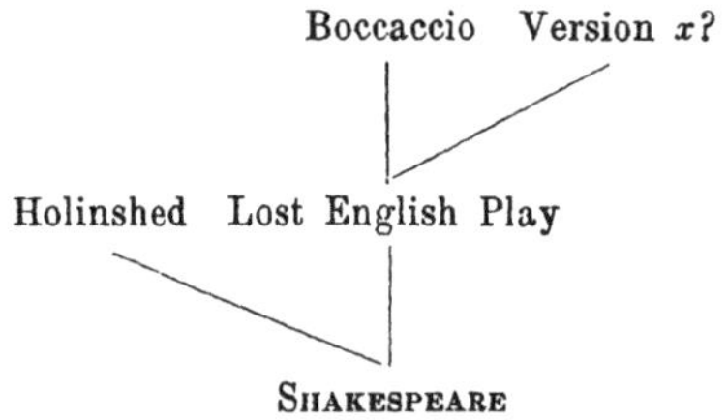

[1] Both Malone and Steevens (Var. Ed., XIII, pp. 2 and 229) name an edition of 1603, which is, however, chimerical. Westward for Smelts was first entered on the Stationers' Registers in Jan. 1620. It has been reprinted by the Percy Society. The Tale told by the Fishwife of Standon-the-Green is reprinted in "Shakespeare's Library". Part I, vol. 2.

[2] I am glad to see my view confirmed by Koeppel, in his Studien zur Geschichte der italienischen Novelle, 1892, p. 73.

[3]) Other critics take a different view. Compare Herford, Eversley Edition of Shakespeare, vol. IV, 113ff.

2. The story of Helena and Bertram in All's Well that Ends Well comes from Boccaccio, Decameron, 3rd Day, 9th Novel. Painter had told it in his Palace of Pleasure, where Shakespeare may have read it.

3. The story of the three caskets in The Merchant of Venice, though occurring in the Decameron (X, 1), has greater affinity with the version in the GESTA ROMANORUM (in the English translation). But probably Shakespeare revised an older play. (See *post.*)

BANDELLO.

1. The Claudio and Hero story in Much Ado About Nothing is in many respects similar to the tale related by Bishop Bandello in his twenty-second novel. The following pedigree, which I tentatively put forward, will, I think, give a clear idea of the relation of the various versions of the story. For particulars I must refer the student to Furness, XII, Dunlop, Shakespeare-Jahrbuch, XXI, 310; XXII, 272; XXXIV, 338, Cohn, Dyce, etc. Holleck-Weithmann, in his work entitled 'Zur Quellenfrage von Shakespeare's Lustspiel Much Ado About Nothing', Kieler Studien, 1902, attempts to prove anew, that Ayrer's 'Schoene Phaenicia' is, in part, ultimately derived from an old English drama used by Shakespeare. H.-Weithmann makes the best of the case. But it is so inherently weak, that all arguments fail to strengthen it.

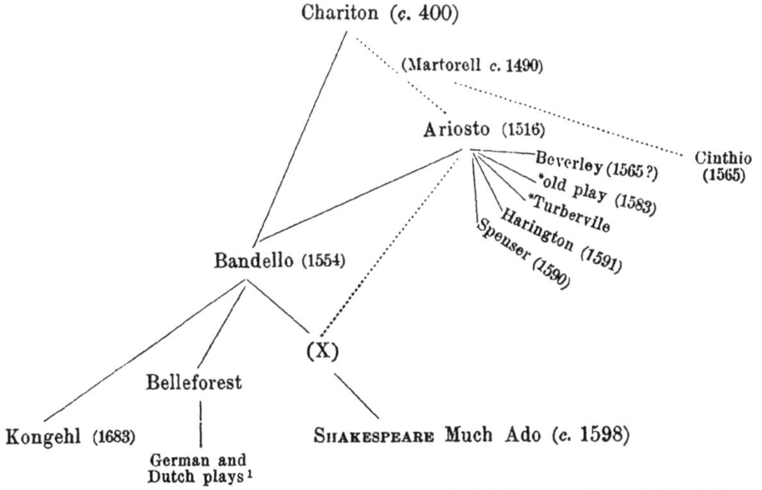

N. B. dotted lines signify: probable influence.
X=hypothetical old play. *=lost

1) By Ayrer (c. 1595); Starter (1618); Kranich (1620); Bährholtz (1674). 'Phoenix (1623).

The points common to both Bandello and Shakespeare, and not to be found in Ariosto, are the following:—the scene laid at Messina; the return from a successful war; the names Lionato de' Lionati and Piero d'Aragona which, with Shakespeare, become Leonato and Pedro of Arragon; the whole later portion of the story after the scene at the window: the supposed death of the lady rejected by her lover; her secret revival; her seclusion; her pretended funeral, with an epitaph on her tomb; repentance of the lover, who is to espouse a lady selected by the injured father, who, of course, chooses his daughter; the double marriage.

With the earlier portion of the story we find some agreement among the three authors. Shakespeare's drama, though derived from Bandello, has adopted one incident from Ariosto's 'Orlando Furioso': the maid's personation of her mistress at the window, to which there is nothing parallel in Bandello, and which is too striking to be attributable to accident. This circumstance could not have been taken from Spenser's Fairy Queen, Bk. II, c. 4, st. 17ff., where, it is true, the maid Pryene personates her mistress; but this does not take place at the window.

2. The reader of Bandello's twenty-first novel of the third volume will be strongly reminded of Aaron's inhuman devilishness in 'Titus Andronicus', The novel had been rendered familiar to Englishmen through a ballad, still extant[1], of which we find the first notice in the Stationers' Registers in 1569—1570 (Arber's Transcripts, I, 406).

GIRALDI CINTHIO.

The seventh novel of the third Decade of Cinthio's *Hecatommithi*[2] is the basis of 'Othello'. I prefer to believe that Shakespeare rehandled an older drama founded on this story.

Regarding Measure for Measure, see *post*, s. v. Whetstone.

[1] Printed in the Old Ballads, 1723, vol. II, 152; reprinted in Roxburghe Ballads (ed. by Hindley. 1873—4) and probably elsewhere. Comp. Koeppel, Englische Studien, XVI, 365ff.

It has been frequently remarked (cp. Malone's Var. Ed., XXI, 258) that the second volume of Painter's Palace of Pleasure contains a reference to Titus Andronicus and Tamora. This is a figment.

[2] Cinthio's novelettes were translated into French by Chapuys. Premier (et deuxième) volume des cent excellentes nouvelles de J. B. Giraldy Cynthien .... contenant plusieurs beaux exemples et notables histoires, etc., mis d'italien en françois, par Gabr. Chapuys. Paris, 1583 ou 1584. 2 vol. (Brunet.) On Othello's sources, compare Hart's edition in the 'Arden Shakespeare', 1903.

SER GIOVANNI FIORENTINO.

1. The bond-story, containing the incident of the pound of flesh, in The Merchant of Venice is most closely related with Giovanni's version of it in his *Il Pecorone*, Giornata IV, Novella 1. The story is also told in the Gesta Romanorum, in the Cursor Mundi, and elsewhere; but it is in Giovanni alone that we find the *motif* of the ring and the name Belmont. But here again we have reason to suppose that Shakespeare's direct source was an older play (see *post*).

2. The story of Bucciolo and Pietro Paolo in the *Pecorone* (Giornata I, Novella 2) may have supplied some hints for The Merry Wives of Windsor. Of this Novel we possess an English version entitled 'Two friends went to study at Bologna in Italy', etc., contained in 'The Fortunate, the Deceived, and the Unfortunate Lovers', printed in 1632. No earlier edition is known. We also find two tales in

STRAPAROLA'S

Le tredeci Piacevoli Notti (Notte II, Favola 4, and Notte IV, Favola 4), which bear resemblance to the incidents of The Merry Wives. The latter was retold by Tarlton in his 'Newes out of Purgatorie', 1590, under the title, 'The Tale of the Two Lovers of Pisa', etc.[1] Probably Shakespeare was under obligation to some other English version now lost. On these and similar tales and their relation to one another, compare Simrock, Die Quellen des Shakespeare.[2]

GL' INGANNATI AND THE SOURCE OF TWELFTH NIGHT, etc.

Gl' Ingannati, an extremely popular Italian play, from which a number of dramas and novels take their origin, must be looked upon as the ultimate source of the serious portion of Twelfth Night. That Shakespeare used some version of the Ingannati cycle is certain from the fact that the prominent features of Gl' Ingannati and the dramas and novels closely following this Italian play reappear in his drama.

[1] See Shakespeare's Library, Part I, vol. III; Old Shakespeare Society, 1844, p. 95; and 1842, Appendix to First Sketch of The Merry Wives, pp. 75, 125; and Dunlop, Hist. of Fiction, 1888, vol. II, p. 159. Straparola was translated into French, 1560 etc.

[2] There is but slight resemblance between The Fishwife's Tale of Brainford, contained in Westward for Smelts, 1620, and The Merry Wives. But the scene is laid in Windsor; and the tale is put into the mouth of a woman of Brainford. On Julian of Brainford, see *post*.

In all of these we have: the brother and sister exactly resembling each other; the latter enamoured of a youth; serving him in disguise of a page; and sent as a messenger to an obdurate lady, who falls in love with the pretty go-between and is betrothed to the brother whom she mistakes for the page; the conclusion with a double marriage. On some version of the Ingannati, probably on Bandello's[1] or Belleforest's novel, Barnabe Riche, in his 'Farewell to Militarie profession', 1581, constructed his narrative of Apolonius and Silla, which, compared with the other versions, appears simplified by the expunction of all the secondary characters. This, I take it, explains the changes and alterations found in his story. With Riche *alone* Shakespeare has in common the following points:—the *duke* had *not* previously loved the disguised girl; the shipwreck; the independent position of the lady (the duke's beloved) owing to a recent death in her family; the lover finally going to the lady's house, by whom he is definitely rejected; the discovery of the apparent double faithlessness of the page accused before the duke. These resemblances cannot possibly be denied. But I believe there are also some features, though not very striking, common to the Ingannati drama and Shakespeare, and wanting in Riche: the brother coming with a companion to whom the town is well known; the inn at which the companion remains while the brother roams about the town *circumspiciendi causā*; a messenger sent by the lady to invite the page mistakes the brother for his disguised sister; the lover's threat to kill both the lady and the page. But Riche's duke, also, threatens to put the page to death.

Where did Riche take the story from? From Belleforest, quoth Mr. Furness, on the score of the following passage in Riche: 'Gentle-'women, accordyng to my promise, I will here, for breveties sake, 'omit to make repetition of the long and dolorous discourse recorded 'by Silla for this sodaine departure of her Apolonius'. Here, according to Mr. Furness, we have a covert reference to the versified grief, the forsaken girl utters throughout four pages or more in Belleforest's novel, who is alone in this 'dolorous discourse'. But we can scarcely draw a certain inference from the passage quoted. I may mention as a coincidence between Riche, Belleforest, and Bandello, that the lover's recognition of the page is preceded by a long discussion. On the other hand, the duke in Riche's novel is informed by his servants of the page's double dealing. This is a trait to be found in the Ingannati

[1] Novel 36 of Part 2.

drama, but not in Bandello and Belleforest. On the whole, Riche's novel appears so altered that it is difficult to decide with absolute certainty where he drew his materials from. But we cannot be far wrong in supposing that he had read Belleforest or Bandello.

It has been supposed that the 'Ingannati' is immediately connected with Shakespeare. Shall we suppose that the Italian play which, we know, gained a footing in France and Spain, was acted in the English tongue on a London stage? Recently, a Latin version of the Ingannati, called 'Laelia', which was performed in Cambridge in 1590 and 1598 has been unearthed. As it does not help us much I pass it by. An account of it will be found in the Shakespeare Jahrbuch, xxxiv, pp. 286, 291. See also Furness, xiii. The variorum editor, who ignores the resemblances between Riche and Shakespeare, missed another track which will perhaps lead us nearer to the goal. Riche, in the conclusion to his 'Farewell to Military Profession', says that some of the stories of his collection had been 'presented on a 'stage'. It is a great pity that he was not more explicit in his statement. Now, we meet with an extraordinary circumstance which demands our attention. There is preserved to us in print a German play entitled "Tugend- und Liebesstreit" of the year 1677. This, or rather an earlier version of it, had evidently been performed by English comedians at Graz in 1608, and at Dresden 1626. What is so remarkable about the play is the fact that *Riche's* 'Apolonius and Silla' is the *unmistakeable* source of this German drama. How are we to explain this? Are we to hypothetize an early English play founded on Riche as the common source of both Shakespeare's Twelfth Night and the Tugend- und Liebesstreit? This supposition, I am free to confess, seems to me extremely plausible. As to this interesting question, see the excellent work of Creizenach, 'Die Schauspiele der englischen Komödianten', 1889, where a reprint of the German play will be found.

Taking into consideration all I have said above, and what has been stated by Furness, Meissner[1], Halliwell, Creizenach, Klein, Hunter, Farmer, etc., to all of whom I must refer for details, we may now draw up the following table:

[1] Programm über Shakespeare's 'Was ihr wollt', Part I, Lyck 1895. Meissner has kindly informed me that he is not in position to continue his discourse.

Trial Pedigree:

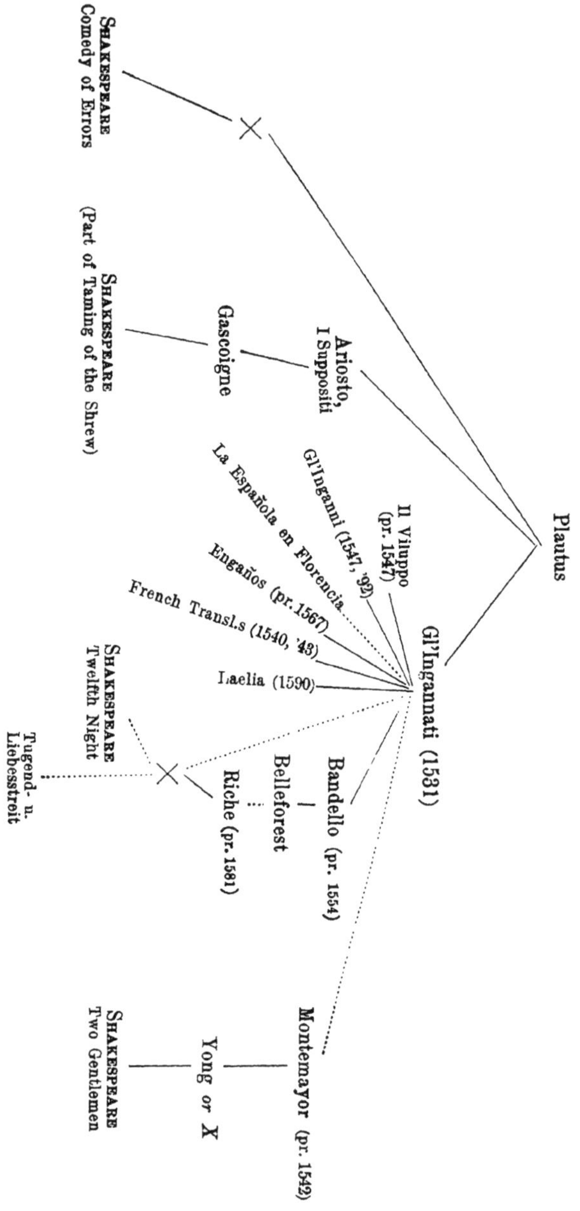

N. B. dotted lines mean: probable, but not absolutely certain, relationship.
×=intermediate link, probably an English drama, now lost.
pr.=printed. x=lost play.

To Secco's and Gonzaga's plays called 'Gl' Inganni', to which Manningham refers in his diary, Shakespeare's Twelfth Night seems to be in no way indebted. (cf. Klein, IV, 806.)

THE ITALIAN DRAMA GENERALLY.

Julius Leopold Klein, the author of the prodigious work 'Geschichte des Drama's', is persuaded that Shakespeare was largely influenced by the Italian drama. (See Th. Ebner's Index-volume to Klein's work, s. v. Shakespeare). However, his arguments fail to carry conviction with them for every reader. Dowden, in his Primer of Shakespeare, p. 65, referring to Armado and Holofernes in Love's Labour's Lost, says: "The braggart soldier and the pedant are characters well known "in Italian comedy[1] and perhaps it was from that quarter that the "hint came to Shakspere." The braggart soldier was also a common figure in the English drama; and in Sidney's masque 'The Lady of May' we find the figure of a pedant. If we may believe Stephen Gosson[2], the continental novels and plays were largely drawn upon by Shakespeare's predecessors, and it is possible that the great poet may have experienced in this way some indirect influence of the Italian drama.

Perhaps I ought to mention some curious coincidences in names pointed out by Hunter and Dr. Garnett. Hunter connects Fabia and Malevolti of *Gl' Ingannati* and its prologue with Fabian and Malvolio of Twelfth Night.[3] Cesario, he thinks, is taken from Gonzaga's *Gl' Inganni*, 1592, where we find the name Cesare, assumed by the lady in disguise; and Orsino from *Il Viluppo.*

Dr. Garnett, in his History of Italian Literature, 1898, p. 229 says:

> The novel by Cinthio himself on which his play [Measure for Measure] is founded was dramatised by Whetstone; but that Shakespeare had seen Cinthio's dramatic version [Epitia] also may be inferred from a minute circumstance. Cinthio's play, not his novel or Whetstone's adaption of it, has a character named Angela, whose name disappears from *Measure for Measure*, but who bequeaths Angelo as that of her brother whom Cinthio calls Juristi, and Whetstone Andrugio.

But even Klein in his Geschichte des Drama's, V, 355, considers this coincidence of no weight.

[1] Comp. what Montaigne says about the pedant of the Italian comedy (*Ess.* I. 24).

[2] Compare *ante*, p. 59—60.

[3] N. B. *Benvolio* is the name of a character in Marlowe's Dr. Faustus and in Romeo and Juliet.

ARIOSTO.

As to an incident in Much Ado taken from the *Orlando Furioso*, see above, s. v. Bandello. Joseph Hunter's supposition that Ariosto's description of a tempest at sea (Canto 41.) inspired the storm-scene and some incidents closely connected therewith in The Tempest has met with no acceptance.[1] Ariosto's *I Suppositi* was known to Shakespeare in Gascoigne's translation (s. *post*).

PETRARCA.

Though we discover many Petrarcan conceits in Shakespeare's sonnets, we can scarcely claim for the latter a direct acquaintance with the Italian poet, who had been imitated *ad nauseam* by a whole army of English sonnetists. In one place ('Romeo', II, IV, 40) our poet openly avows his knowledge concerning Petrarca's sonnets, but it does not follow from the passage, that he had read them. Mercutio, speaking of Romeo, says:

> Now is he for the numbers that *Petrarch* flowed in: *Laura* to his lady was but a kitchen-wench; marry, she had a better love to be-rhyme her.[2]

SPANISH LITERATURE.

We are fairly safe in asserting that our poet was unacquainted with the Spanish language. Of one author, however, we find traces in his works,—of

JORGE DE MONTEMAYOR,

whose story of the shepherdess Felismena, in the pastoral romance *La Diana*, furnished the materials for the adventures of Julia and Proteus in The Two Gentlemen of Verona. The first complete translation of the Diana, by Bartholomew Yong,[3] appeared in 1598, but

[1] Dr. Schoembs wrote a Dissertation on the influence of Ariosto's Orlando Furioso on Elizab. Liter. 1898.

[2] I may mention that Petrarca had been translated into French about the middle of the sixteenth century. The *Triumphes* of Petrarch were translated into English by Lord Morley, c. 1554, (reprinted by the Roxburghe Club. 1887).

With Giordano Bruno Shakespeare is in no wise connected. See Shakespeare Jahrbuch XXVI, 258ff., and Robertson 'Montaigne and Shakspere', p. 82ff.

[3] Certain portions of the Diana were also translated by others. "In 1596, "while sojourning in Italy and Germany, Wilson translated from the Spanish "Gorge (*sic.*) de Montemayor's 'Diana'." (Dict. of Nat. Biogr., s. v. Sir Thomas Wilson.) But this was not printed. One Edward Paston, Yong tells us, had 'for 'his owne pleasure ...... aptly turned out of Spanish into English some leaves 'that liked him best.'

it existed in manuscript as early as 1582 or 1583,[1] and part of the romance had been dramatized in 1584 (see *post).*

In Shakespeare Jahrbuch, vol. XXXIV, Dr. Tobler *jun.* has pointed out correspondences between the Diana and Midsummer Night's Dream. The resemblances are perhaps still greater between this play and Alvise Pasqualigo's *Gl' Intricati* (1581) in which incidents of the *Diana* are dramatized. See Dr. Vollhardt's Programme, 'Die Beziehungen des Sommernachtstraumes zum italienischen Schäferdrama'. Leipzig. 1899.[2]

[1] Ward, Engl. Dram. Lit., vol. II, p. 80. Cp. also Dict. of Nat. Biogr., s. v. Young.

[2] Other programmes and dissertations on Midsummer Night's Dream I need not mention. John Lyly's connexions with the Italian pastoral drama are discussed by Mr. Bond in his edition of Lyly, vol. II, pp. 473 ff.

In Dowden's edition of Cymbeline which has just appeared (Arden Shakespeare) an episode in Tasso's Gerusalemme Liberata Bk. VII (translated by Fairfax in 1600) is suggested as a *possible* source of Imogen's adventures at Belarius's cave. (I mention this here for want of space above.)

APPENDIX TO CHAPTER 2.

THREE CHIMERICAL SOURCES.

I—. Diego Hurtado de Mendoza.

Mendoza's *Lazarillo de Tormes,* the first *picaresque* romance of modern literature, which seems to have enjoyed considerable popularity, insipid though it now appears to us,[1] is by some commentators (Eschenburg was the first to make the suggestion) supposed to be alluded to in Much Ado about Nothing, II, i, 205-6: —

> Ho! now you strike like the blind man: 'twas the boy that stole your meat, and you'll beat the post.

The incident supposed to be alluded to is this. Mendoza relates that Lazarillo, when a boy, became a blind man's guide, whose sausage he once stole, for which he was severely punished. In revenge, the boy caused the blind man to knock his head violently against a stone pillar while jumping across a supposed gutter.

The romance, it is true, would have been easily accessible to Shakespeare. Having first appeared in Antwerp in 1553, it was frequently reprinted. Several translations into English were published. The Stationers' Registers record the license of an English version as early as 156$\frac{8}{9}$ (Arber, I, 378). The earliest extant translation is dated 1576. Other editions were printed in 1586 and 1596. (Compare Hazlitt's Bibliography, and Lowndes).

There is therefore no external reason against the above supposition. But let us not forget the chief question. Does the passage referred to contain an allusion at all? Dr. Martin Luther's principle, that the *first* meaning of a bible passage should be considered as decisive is a very healthy one, which Shakespearean scholars have sometimes need of laying to heart. Let us apply this principle. From the context it is clear that Benedick desires Claudio to understand, that he has not deserved his ill-

---

[1] The book is easily accessible in the German translation, published by Reclam, Leipzig, No. 1389. Price 2d.

will. 'Claudio', he means, 'you vent your displeasure on the wrong object. 'You behave like the blind man who, instead of beating the guilty boy, 'who stole his meat (given to him as an alms), strikes wide of the mark, 'and hits the post (which would be the object nearest to him sitting 'before the door of a house as a beggar)'.

If this interpretation be allowed to be correct, there is no room for our friend Lazarillo. Mr. Furness, while disposed to reject Eschenburg's supposition, thinks, nevertheless, that the expression "*the* blind man" refers to some definite *anecdote.* I do not think it does so, necessarily; though this is possible. (On this question concerning Lazarillo, compare Shakespeare Jahrbuch, VI, 353; Furness, XII, 77; Brandl, Schlegel-Tiek Translation, VIII, 236.)

II—· Another Mare's Nest.

Espejo de Principe y Cavalleros is a Spanish Romance by Diego Ortuñez de Calahorra, translated into English in Shakespeare's days under the title of The Mirror of Princely Deeds and Knighthood. Steevens suggested that Falstaff's words: "Phoebus, he, that wandering knight so "fair" (1. Henry IV., Act 1, II, 16) contain an allusion to this work or some ballad founded thereon. But, though so distinguished scholars as Dyce, Aldis Wright, etc. concur with Steevens, I cannot but demur to this view, which seems to me extremely hazardous. Why should not the Sun be likened to a knight-errant,—an idea which would seem natural and simple enough, especially for those days, when the Sun was reckoned a planet or wandering star? (The above mentioned work, by the way, seems to have been confounded with Villalumbrales's Cavallero del Sol.—See Brunet.) Douce, in his Illustrations of Shakespeare, I, 415, advanced an equally untenable supposition.

III—· Antonio de Eslava.

Edmund Dorer, in 'Das Magazin für die Litteratur des In- und Auslandes', 31. Jan., 1885, draws attention to a narrative which has greater affinity with the plot of The Tempest than any tale yet discovered. It is contained in a Spanish collection of stories entitled 'Las noches de invierno por Antonio de Eslava, 1609'. The narrative, which may, like others of the collection, be based on an older tale, is briefly this:

> Dardanus, King of Bulgaria, a virtuous magician, is dethroned by Niciphorus, Emperor of Greece, and has to flee with his only daughter, Seraphina. They go on board a little ship. In mid ocean Dardanus, having parted the waters, rears by art of magic a beautiful submarine palace, where he resides with his daughter till the time she

becomes marriageable. Now the father, in disguise of a fisherman, carries off the son of Niciphorus to his palace under the sea. Of course, the youth falls hopelessly in love with the maiden. The emperor having died in the meantime, Dardanus returns with his daughter and his son-in-law to his former kingdom, which he leaves the latter to rule over, while he withdraws into solitude.

The resemblances are obvious. But it is by no means certain that the tale is Shakespeare's source, as Dorer thinks.

CHAPTER 3.

THE ENGLISH NON-DRAMATIC POLITE LITERATURE.

SECTION 1.—PRE-ELIZABETHAN AUTHORS.

GEOFFREY CHAUCER.

THE list of pre-Elizabethan poets is worthily opened by that 'morning 'star of song',

> 'Dan Chaucer, the first warbler, whose sweet breath
> 'Preluded those melodious bursts, that fill
> 'The spacious times of great Elizabeth
> 'With sounds that echo still.'

I know of three articles dealing with Shakespeare's relation to Chaucer, but all of them incomplete.[1] My own notes as well as these articles supply the groundwork of the following remarks. To

THE CANTERBURY TALES,

in general, there is an apparent allusion in 'Lucrece', 790:

> And fellowship in woe doth woe assuage,
> As palmers' chat makes short their pilgrimage.

THE KNIGHT'S TALE,

we may take for granted, Shakespeare had read. A Midsummer Night's Dream certainly shows some points of agreement with the Theseus-Hippolyta part of the story. North's Plutarch, which Shakespeare

[1] Ballmann, Anglia XXV, 1902; Sarrazin, Anglia Beibl. VII, 265, who gives some more or less doubtful parallelisms. This parallelism work of Sarrazin is very dangerous. Poole, in his Index to Period. Literat., mentions an article on Shakesp. and Chaucer in the Quart. Rev. vol. 134 (1873), p. 225. But this does not help us much.

demonstrably did use for his Mids. N. Dr., would have supplied a good many hints for the Thesean framework. But not all. Philostrate does not occur in Plutarch, nor does Plutarch make mention of doing observance to May. Theseus is not called *duke* of Athens by him. These traits at least, as well as the Athenian wood, and the romantic costume generally, in which the classical characters appear, are derived from Chaucer. But another problem presents itself. Had Shakespeare seen the Knight's Tale acted on the stage? There is some reason to suppose he had. And, what is more to our purpose, the play has apparently left its traces on our Mids. N. Dream. Richard Edwardes had dramatized the story in 1566. This play of his, 'Palamon and Arcite', was acted before the Queen in Oxford in the same year. Detailed accounts of the play (now lost) and its performance are preserved to us in contemporary MS. reports of Queen Elizabeth's visit to Oxford, published partly by Nichols (Progresses of Q. Eliz., 2nd ed.) and partly by Plummer (Elizabethan Oxford, 1887). The name Philostrate, I may remark, which has been claimed as a striking proof for Shakespeare's use of Chaucer, was in that play. Theseus was the 'dux Athenarum', according to a Latin account.[1] But what is more important, 'in the said play was acted *a cry of hounds* in the Qua-'drant, upon the train of a fox *in the hunting of Theseus.*'[2] This part, we are distinctly told, proved extremely popular and successful. The same scene was repeated in Oxford on another occasion in 1583 (Nich., II, 409) and again acted or imitated before the Queen in 1572 (cp. Malone, III, 369, 'Hunters'). Thus, I conclude, it came about, owing to theatrical tradition, that Shakespeare introduced into his play (Act IV, sc. I) the hunting of Theseus and the music of the hounds, which was probably *really* mimicked behind the scenes.[3] No doubt Edwardes's play was performed on the London stage. What relation The Two Noble Kinsmen bears to it and to the 'Palamon and Arsett', mentioned as a play by Henslowe in 1594, it is impossible to say with certainty.

From

THE MERCHANT'S TALE.

Shakespeare is said to have borrowed the *motif* of the quarrel among supernatural beings (Pluto *vers.* Proserpine—Oberon *vers.* Titania).

[1] Plummer, p. 128.

[2] Nichols, I, 212.

[3] "A cry of hounds, and horns winded in peal" is a stage-direction in Titus Andron., II, II; Cp., too, I, II, 494.

But Pluto and Proserpine do not quarrel. They debate in perfect friendship. As You Like It, IV, I, ll. 159—179 have been appositely compared with Merch. Tale (E. 2264—2275).

THE HOUSE OF FAME.

The emperor's court is like the house of Fame,
The palace full of tongues, of eyes, and ears.
(Tit. Andron. II, I, 126.)

We have here a reference to Chaucer's famous poem. We do, indeed, find a description of Fama's abode in Ovid (Metam. XII, 39—64); But the expression 'house of Fame' is not in Golding's Translation. Moreover, the idea of the many tongues, eyes, and ears, not in Ovid, must be due to Chaucer, whose Fame had 'as fele iyen As fethers 'upon foules be ..... also she Had also fele up standyng eares And 'tonges, as on beast been heares' (H. of F. III, 291f.).[1] Though Virgil describes Fame in a similar manner in Aen., IV, 173ff., (see esp. 181—183), he says nothing of her dwelling.

TROILUS AND CRISEYDE.

Shakespeare used this poem as the main source of his Troilus story [Troil. I, I & II; III, I & II; IV, I—IV & V (12—63); V, II & III (95 – end)].[2] For particulars see the late Dr. Small's scholarly work: The Stage Quarrel, p. 154ff. Shakespeare alludes to the story in Ado, V, II, 31; All's Well, II, I, 100; As Y. Like It, IV, I, 97; Lucr., 1486; Merch., V,I, 4—6; Tw. Night, III, I, 57—62; Shrew, IV, I, 153.

In Tw. Night, III, I, 62 we read: "Cressida was a beggar"; and in Henry V., Act II, I, 78—80:

to the spital go,
And from the powdering-tub of infamy
Fetch forth the lazar kite of Cressid's kind.

Nothing to parallel these two passages will be found in our modern editions of Chaucer.[3] But Shakespeare found something in *his*. In Stowe's edition 1561 (and in editions before and after) 'Troilus and

[1] I am quoting from Stowe's edition, 1561, which Shakespeare most likely used. In and after 1598 he may have used the next edition after Stowe's, by Speght, 1598 and again 1602. Cp. Skeat's ed., Introd., vol. I (Rom. of the Rose, etc.) p. 27f.

[2] The story had been dramatised in 1599 by Dekker and Chettle, as we know from Henslowe's Diary.

[3] Of course, I do not forget that Skeat's edition includes a volume of doubtful and spurious pieces.

Criseyde' is immediately followed by The 'testament of Creseide' (composed by Robert Henryson, as we know), as though it were Chaucer's continuation of the former poem. In this pseudo-Chaucerian piece Creseyd meets with her due reward for her unfaithfulness, is delivered to the spital-house, and dies a leprous beggar. "Kit(e) of "Cressid's kind" seems to have been an almost proverbial expression in Shakespeare's days.

The verses in Chaucer's Troilus, III, 1422 f., compared with the morning scene of the parting of Romeo and Juliet, deserve the special attention of the Shakespearean student.[1]

THE LEGEND OF GOOD WOMEN.

1. Pyramus and Thisbe. This story, popular in Shakespeare's days, was probably known to him in more than one version, no doubt also in the Chaucerian. Ballmann[2] says it had been dramatized before the date of Mids. N. Dream. But there is not the slightest foundation for this allegation. (Cp. above, Ovid.)

2. On the Legend of Lucretia, see under Ovid.

A PSEUDO-CHAUCERIAN 'PROPHECY',

to be found in some earlier editions of Chaucer (among them, Stowe's ed., 1561, which I have seen), and quoted by Puttenham in his 'Arte of English Poesie' 1589 (ed. Arber, p. 232), is apparently imitated by the Fool in King Lear (III, ii, 80—end). The concluding couplet:

Then shall the Realme of Albion
Be brought to great confusion,[3]

is quoted by the Fool, with the substitution of 'come' for 'be brought'.

To moral

GOWER'S

Tale of Florent there is a clear reference in The Taming of the Shrew, I, ii, 69:

Be she as foul as was Florentius' love,

---

[1] Do not believe what Fränkel says in his laboured treatise, Shakespeare und das Tagelied. 1893.

[2] *ut sup.*, p. 8.

[3] I am quoting from Puttenham, who may have had a version before him differing from that in Stowe's ed. or that reproduced by Skeat (Chauc. ed., I, pp. 45—46). For he agrees with Shakespeare in reading 'realm', instead of 'lond' of the latter versions. But Puttenham may be Shakespeare's source.

The story is in the CONFESSIO AMANTIS, Bk. I, 1407 ff. The name of the hero, Florent, is 'probably due to Gower, who is apt to attach to 'his stories names of his own choosing'.[1] In the first marginal note to this tale we find 'Florencius', as the Latin equivalent for this name.

We find no further certain traces of Gower except in 'Pericles' which is only in part Shakespeare's. The sources of the play are twofold: GOWER'S *Confessio Amantis*, Bk. VIII, ll. 271—2008, and LAURENCE TWINE'S *Patterne of painefull Adventures* (entered on the Stat. Reg. 1576, see Arber, Transcr. II, 301—; printed without date, and again in 1607). For acts 1—3 Gower is the main source and Twine rarely made use of. For Acts 4 and 5 the case is the reverse, Twine being here the chief source. For a full discussion see E. Klebs, 'Apollonius aus Tyrus', Berlin, 1899, pp. 474 ff. George Wilkins's novel, 'The Painfull Adventures of Pericles Prince of Tyre', 1608, is posterior to the drama, to which it refers on the title-page as well as in the Argument.[2]

WILLIAM CAXTON'S

Recuyell of the Historyes of Troye, the first book printed in the English language, 1474?, frequently republished in the 16th and 17th centuries—Hazlitt mentions an edition of 1596[3], which Shakespeare may have used—supplied most of the incidents of the Trojan war for 'Troilus and Cressida'. Lydgate's Troyebook was not made use of (see Small, The Stage-quarrel, pp. 156 ff.; Cp., too, Malone, VIII, 455).

There is also some connexion (direct or indirect) between Shakespeare and the second English book printed, that is: Caxton's *Game and Playe of the Chesse* (first issued in 1475?), where we find the following quaint etymology (Bk. 3. ch. 5):—

> For the women ben likened unto softe waxe or softe ayer and therfor she is callid mulier whyche Is as moche to saye in latyn as mollys aer. And in english soyfte ayer. And it happeth ofte tymes that the nature of them

---

[1] Gower, ed. Macaulay, 1901, note *ad loc.*

[2] Smyth, in 'Shakespeare's Pericles and Apollonius of Tyre', Philadelphia, 1898 (p 63), repeats the old legend of an earlier play 'alluded' to in the Alleyn papers. The entry was forged by Collier (see 2. Quarto Facsimile by Praetorius with an Introd. by Round, p. III). Smyth (p. 60) also gives an unauthenticated date, 1596, for 'Pimlyco', where the play is mentioned. This date is chimerical. (Cp., too, Malone XXI, p. 4).

[3] Entitled: 'The Auncient Historie, of the destruction of Troy'. Hazlitt, Handb. *s. v.* Fevre.

that ben softe and mole taketh sonner Impression than the nature of men that is rude and stronge.

Compare 'Cymbeline', V, iv, 140; V, v, 437; and, especially, the following lines 446ff.:

The piece of tender air, thy virtuous daughter,
Which we call 'mollis aer'; and 'mollis aer'
We term it 'mulier': which 'mulier' I divine
Is this most constant wife .....

SECTION 2.

ELIZABETHAN AUTHORS.

ARTHUR BROOKE'S

Tragicall Historye of Romeus and Iuliet[1] is the basis on which Shakespeare built his wonderful drama, Romeo and Juliet. (The source-question has been carefully and minutely examined by Mr. P. A. Daniel, in the New Shaksp. Soc., III, 1, pp. XIIff.; and by Dr. Schulze in Sh. Jahrb., XI, 195f., and 218ff.[2])

The late Prof. Zupitza[3] pointed out some *motifs* and incidents in The Two Gentlemen of Verona, which are traceable to Brooke's poem. Silvia refuses to accept the hand of Thurio, though favoured by her father, and decides to marry Valentine secretly. For the purpose of escaping with her, Valentine undertakes to "climb her window, "The ladder made of cords" (II, iv, 181). He is banished and flees

[1] First printed by R. Tottell in 1562. 'Romeo and Juletta' was licensed to Tottell in 1583 (Arber, Stat. Reg., II, 419). This is no doubt Brooke's poem. But no copy of this edition, if published, is known. Brooke's poem was reissued by Robert Robinson in 1587.

[2] Regarding the history and development of the Romeo-story, compare, too, Fränkel's papers in Zeitschrift für vergl. Literaturgesch. Neue Folge, III & IV, and in Kölbing's Engl. Studien, XIX, 183. Regarding a Latin drama, but which is later than Sh.'s play, see Sh. Jahrb., 34, p. 255. A terrible amount of learning has accumulated round Shakespeare's play.

[3] See Sh. Jahrb. XXIII, pp. 1—17. Here will be found a careful examination of the sources of The Two Gentlemen of Verona. I cannot, however, endorse Zupitza's supposition, which he shares with some other *littérateurs*, that a hypothetical old English play, the alleged original of 'Julius und Hyppolita' (a German play printed in the collection 'Engelische Comedien und Tragedien', 1620, and reprinted by Cohn, 'Sh. in Germany', and discussed by Creizenach, Die Schauspiele der englischen Komoedianten, 1889, p. LIV), is one of Shakespeare's sources.

towards Mantua. Silvia's father is wroth with her. Under the pretext of making confession at Patrick's cell, she escapes, noticed only by Friar Laurence. The names Verona and Mercatio (for Mercutio) may be derived from the same source. The sudden change of affection in Proteus is analogous to Romeus's sudden change, the instant he sees Juliet:

His [Romeus's] former love, for which of late he ready was to dye,
Is nowe as quite forgotte, as it had never been:
The proverbe saith unminded oft are they that are unseene.
And as out of a planke a nayle a nayle doth drive:
So novell love out of the minde the auncient love doth rive.
(Brooke, vv. 204—'8)

The same simile is used by Shakespeare in The Two Gentlemen, II, IV, where the following verses are put into the mouth of Proteus, after he had seen Silvia for the first time:

Even as one heat another heat expels,
Or as one nail by strength drives out another,
So the remembrance of my former love
Is by a newer object quite forgotten. (ll. 192—195.)

(Comp. Coriol., IV, VII, 54:

"One fire drives out one fire; one nail, one nail".)

There are two or three more traces of Brooke in Shakespeare: The name *Escalus* is given to the good Lord in Measure for Measure, and occurs also once in All's Well. In this latter play Diana is made to say: "I am, my lord, a wretched Florentine, Derived from the "ancient *Capilet*" (V, III, 158); cf. "Diana *Capilet*" (l. 147, ib.). The name, I ought to remark, is spelt Capilet or Capelet by Brooke, Capellet by Painter.

Was Brooke the only source Shakespeare used for his 'Romeo and Juliet'?

For convenience' sake I first reproduce here Fränkel's table[1], giving at one glance a clear idea of the literary history and development of the Romeo-story.

[1] Zeitschr., *ut sup.*, III, p. 182.—Mr. P. A. Daniel mentions to me an Italian Romeo-play, La Donna Constante by Raffaello Borghini (1578), which, he says, has so far remained unnoticed.

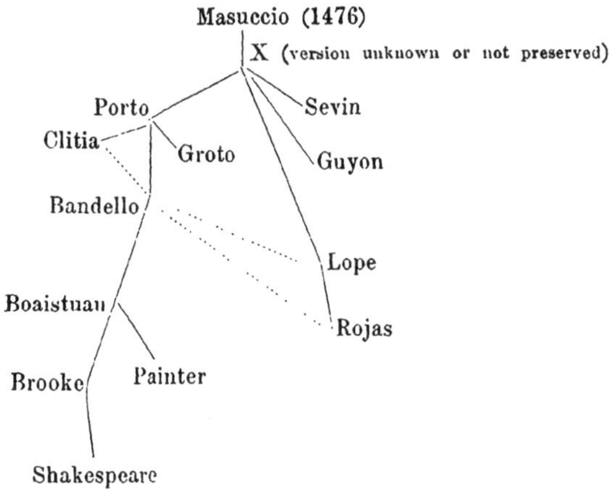

N. B. dotted lines signify probable secondary influence.

If we suppose, that Shakespeare's knowledge of French and Italian was such that he preferred to read English translations instead of their originals, it will be natural to seek for *English versions* of the Romeo-story. Apart from Brooke's poem, the only other pre-Shakespearean English version preserved to us is that given by Painter in the second volume of his 'Palace of Pleasure' (pr. 1567 and again *circa* 1575).[1] This novel of 'Rhomeo and Julietta' translated from Boaistuau has been subjected to close comparison with Shakespeare and Brooke since Steevens's and Malone's days, a comparison which irrefragably establishes, that Brooke is Shakespeare's source, as I have already said.

In Shakespeare Jahrb., XI, p. 218, two or three trifling coincidences between Shakespeare's play and Painter's novel are given. 1) While Brooke has 'Romeus', Painter has Rhomeo. Romeo is the Italian form of the name. 2) Shakespeare's Romeo gives the apothecary forty *ducats* for the poison (V, I, 59); Painter has 'Fifty *Ducates*; Brooke: 'fiftie *crownes of gold*'. But Shakespeare knew that ducats were current in Italy (cp. Merch. of Ven.) 3) Act IV, I, 105: "Thou shalt "continue *two and forty hours*." 'Brooke does not mention the time 'which the sleeping draught is to hold Juliet. Steevens notes as proof 'that Shakespeare consulted Painter, that in Painter it is said to be

[1] I may remark here, that Warton's supposition (repeated by Drake and Fränkel) of an English translation of Boaistuau and Belleforest's novels in 1596 is chimerical. Warton was no doubt misled by an entry in the Stat. Reg., III, 67, which, however, refers to Silvayn's 'Orator'.

'"*40 houres at the least*". On this Boswell remarks (Var. Ed. VI, p. 265), '"although the number of hours ..... are not specified in the poem, 'yet enough is said to make it easily inferred, when we are told that 'two [?] nights after, the Friar and Romeo were to repair to the se-'pulchre". Da Porto has *forty-eight* hours; Clitia, *two days*; Bandello 'and Boaistuau *about forty hours;* Groto, in 'La Hadriana', *about sixteen 'hours.*' (Daniel, *ut sup.*, p. XVI). Boswell's explanation, I must confess, does not quite satisfy me. I cannot infer the forty-two hours from Brooke's poem. 4) To these three points I add a fourth. '*Anselme*' is one of the invited guests (I, II, 68). Now, Anselme is the name of the messenger sent by Friar Lawrence to Mantua, in Painter. By Brooke and Shakespeare he is called Friar John.—Seeing that these coincidences are not very striking, we have to conclude that the probability of Shakespeare having used Painter is not great.

Regarding the question of a pre-Shakespearean play on this theme see *post*, in the chapter on the Drama.

'Resemblances between passages of Shakespeare's tragedy and 'passages of Groto's Italian tragedy of *Hadriana* are probably due to 'accident.' (Cp. Daniel, *ut sup.* p. XXIff.; and Sh. Jahrb., XI, 196ff).

SAMUEL DANIEL.[1]

1. The Complaint of Rosamond

is important in its bearing on Shakespeare, in whose earlier works we find remarkable reminiscences, in the language, substance and form, of that poem. Daniel's 'Rosamond' was first printed together with his 'Delia' in 1592. The poems met with the applause of the public and ran through another edition in this year and a third (and fourth?) in 1594.[2]

To begin with Shakespeare's *epic poems*, 'A Lover's Complaint' is written in imitation of Daniel's poem, as the subject, characterization, tone, metre, and style show. This must be felt. The main motive of each poem is the loss of virginity by the heroine, into whose mouth the complaints are put. A similar topic treated in a somewhat similar way is The Rape of Lucrece. Thomas Churchyard's 'Shore's Wife', the most popular poem in the Mirror for Magistrates, and much akin to 'Rosamond', may also have been known to Shakespeare, though I cannot find any clear traces of it in his works.

[1] Besides notes of my own, I have made use of Dr. Ewig's article in Anglia XXII, p. 436—448, to which I refer the curious reader.

[2] See Grosart's ed., 1885, I. pp. 20f. and 80.

The following parallelisms in Lucrece and the Complaint of Rosamond are very striking.

I) Ros. 128.

> Ah *beauty* Syren, faire enchaunting good,
> Sweet *silent Rhetorique of perswading eyes:*
> Dombe Eloquence, whose powre doth move the bloud,
> More then the words or wisedome of the wise.

Compare Lucr., 29—30:

> Beauty itself doth of itself persuade
> The eyes of men without an orator.

Comp. also Love's Lab. Lost, IV, III, 60:

> The heavenly rhetoric of thine eye.

II) Ros. 246:

> Thou must not thinke thy flower can alwayes flourish,
> And that thy *beauty* will be still admired;
> But that those raies which all these flames doe nourish,
> *Cancell'd* with Time, will have their *date expired.*

Compare Lucr. 22ff.:

> O happiness enjoy'd but of a few!
> And, if possess'd, as soon decay'd and done . . .
> As is the morning's silver melting dew
> Against the golden splendour of the sun!
> An *expired date, cancell'd* ere well begun:
> Honour and *beauty,* in the owner's arms,
> Are weakly fortress'd from a world of harms.

III) Ros. 439:

> Com'd was the *Night, mother of sleepe and feare,*
> Who with her *sable*-mantle friendly covers
> The sweet-stolne sport of ioyfull meeting Lovers.

Compare Lucr. 117:

> Till *sable Night, mother of Dread and Fear,*
> Upon the world dim darkness doth display,
> And in her vaulty prison stows the Day.

The situation in both poems is, *mutatis mutandis,* similar.

IV) The following parallel passages exemplify the structural resemblance in an obvious manner:

Ros. 428ff.:

> I saw the sinne wherein my foote was entring,
> I saw how that dishonour did attend it,

I saw the shame whereon my flesh was ventring,
Yet had I not the power for to defend it.
So weake is sence, when error hath condemn'd it.
  We see what's good, and thereto we consent,
  But yet we choose the worst, and soone repent.

Lucr. 491f.:

I see what crosses my attempt will bring;
I know what thorns the growing rose defends;
I think the honey guarded with a sting;
All this beforehand counsel comprehends:
But will is deaf and hears no heedful friends;
  Only he hath an eye to gaze on beauty,
  And dotes on what he looks, 'gainst law or duty.

We find also traces of Daniel's Rosamond in Shakespeare's *dramas*, notably in Romeo and Juliet. The King's complaint in 'Rosamond' over the death of his love is distinctly echoed in Romeo's speech which he utters on seeing Juliet apparently dead in the tomb.

Ros. 773:

And nought-respecting *death* (the last of paines)
Plac'd his *pale colours (th' ensigne* of his might)
Upon his new-got spoyle before his right.

Ros. 834—847:

Pittifull mouth (saith he) that living gavest
The sweetest comfort that my soule could wish:
O be it lawfull now, that dead thou havest,
This sorrowing farewell of a dying kisse;
And you faire eyes, containers of my blisse,
  Motives of Love, borne to be matched never,
  Entomb'd in your sweet circles, sleepe for ever.

Ah, how me thinkes I see Death dallying seekes,
To entertaine it selfe in Loves sweet place;
Decayed Roses of discoloured cheekes,
Doe yet retaine deere notes of former grace:
And ugly Death sits faire within her face;
  Sweet remnants resting of Vermillian red,
  That Death it selfe doubts whether she be dead.

Compare Romeo, V, III, 91—96; 101—105; 112—115:

O my love! my wife!
Death, that hath suck'd the honey of thy breath,
Hath had no power yet upon thy beauty:
Thou art not conquer'd; beauty's ensign yet

Is crimson in thy lips and in thy cheeks,
And death's pale flag is not advanced there ....

Ah, dear Juliet,
Why art thou yet so fair? shall I believe
That unsubstantial death is amorous,
And that the lean abhorred monster keeps
Thee here in dark to be his paramour? ....

Eyes, look your last!
Arms, take your last embrace, and lips, O you
The doors of breath, seal, with a righteous kiss
A dateless bargain to engrossing death.

In 2. Henry IV., Act V, sc. v, 50, Falstaff uses the exclamation: "My king! my Jove!" which is possibly a reminiscence of Rosam., 239:

Doost thou not see, how that *thy King (thy Iove)*
Lightens forth glory on thy darke estate.

2. Sonnets to Delia,

greatly admired by the Elizabethans and, among them, by Shakespeare, served as a model for his sonnetic flights. But apart from Shakespeare's sonnets, whose sources I do not desire to discuss, except in general terms, we find traces of 'Delia' in his other works.

I) Daniel, speaking of his lady's disdain, says in his 5th Sonnet:

Which turn'd my sport into a Harts dispaire,
Which still is chac'd, while I have any breath,
By mine owne thoughts, set on me by my Faire:
My thoughts (like Houndes) pursue me to my death.

The same image clothed in nearly the same words is in Twelfth Night, I, I, 19ff: —

O, when mine eyes did see Olivia first,
Methought she purged the air of pestilence!
That instant was I turn'd into a hart;
And my desires, like fell and cruel hounds,
E'er since pursue me.

II) Sonnet 48:

My Delia hath the waters of mine eies ....
[which attend her as duly as the ocean the moon]
Yet nought the rocke of that hard heart can move,
Where beat these teares with zeale, and fury drives.

Compare Lucr. 587 ff.:

> If ever man were moved with woman's moans,
> Be moved with my tears, my sighs, my groans:
> All which together, like a troubled ocean,
> Beat at thy rocky and wreck-threatening heart,
> To soften it with their continual motion.

III) With Lucrece', ll. 22—28, quoted above, compare Daniel's 50th Sonnet:

> Beautie (sweet Love) is like the morning dew,
> Whose short refresh upon the tender greene:
> Cheeres for a time, but till the Sunne doth shew,
> And straight tis gone as it had never beene.
> Soone doth it fade that makes the fairest florish,
> Short is the glory of the blushing Rose.

Dr. Ewig also quotes, in illustration of Lucr. 26, from the same sonnet ll. 11, 12:

> .... in Beauties lease expir'd appeares
> The date of Age.

The gentleman does not seem to have been aware of the fact, that these words do not appear in the earlier editions of Delia.

3. The Civil Wars,

four Books of which appeared in 1595, may have been consulted by Shakespeare, when he penned Henry IV. Compare stanza 84, Bk. IV.— King Henry IV. is ill, pain and grief

> Besieg'd the Hold, that could not long defend ....
> Wearing the wall so thin that now the mind
> Might well look thorow, and his [=its] frailty finde'—

which is much like 2. Henry IV., Act IV, IV, 117 ff:

> he [Henry] cannot long hold out these pangs:
> The incessant care and labour of his mind
> Hath wrought the mure that should confine it in
> So thin that life looks through and will break out.

In conclusion I mention, that Daniel had written a play, 'The Tragedie of Cleopatra' (pr. 1594 etc.), as a companion to the Countess of Pembroke's 'Antonie' (translated from the French of Garnier), pr. 1592. To neither of these 'reading' plays is Shakespeare at all indebted for his Antony and Cleopatra'.

EDMUND SPENSER.

With Spenser our poet has very few points of contact. In Mids. N. Dream there is said to be a refererence to THE TEARS OF THE MUSES (1591), in which the nine Muses lament the decline of learning and knowledge.

> The thrice three Muses mourning for the death
> Of Learning, late deceased in beggary—

is the theme of one of the proposed entertainments in Mids. N. Dr. (Act V, I, 52—3). Theseus characterizes it as follows:

> That is some satire, keen and critical,
> Not sorting with a nuptial ceremony.[1]

To Spenser's relation of the Lear-story in THE FAIRY QUEEN (Bk. II, Canto X) Shakespeare seems to be indebted for the name Cordelia, which Spenser evidently invented, because he was in want of a rhyme, and which occurs nowhere else except in the ballad, which is no doubt later than Shakespeare's play. In the other versions she is called: Cordell, Cordila, Cordella. Shakespeare is also said to have taken the idea of Cordelia's death by *hanging* from the Fairy Queen, where she *hangs* herself owing to the necessity of a rhyme again.

Some parallelisms of thought and expression have been pointed out between Shakespeare and Spenser. But the resemblances are not great enough to allow us to infer anything definite. The only more striking one appears to me to be the following:

> But whenas *Morpheus had with leaden mace*
> Arrested all that courtly company—
> (F. Q., Bk. I, IV, 44)

Comp. 'Caesar', Act IV, III, 267:—

> O murderous slumber,
> Lay'st thou thy leaden mace upon my boy?

Was the leaden mace considered to be a mythological attribute of Morpheus?

MARLOWE'S HERO AND LEANDER

is quoted in As You Like It, III, v, 82,

> Dead shepherd, now I find thy saw of might,
> *Who ever loved that loved not at first sight?*

This poem of Marlowe's, the dead shepherd, from which the second line (=Hero and Leander, I, 176) is taken, was a great favourite with

[1] Dr. Garnett refers me to his remarks in 'Literature', Oct. 22, 1898, p. 375.

Shakespeare; and he must have known it by heart. Not only do we find frequent reminiscences of it in his dramas, especially those of the earlier period, but the two poems 'Venus and Adonis' and 'Lucrece' drew much of their inspiration from it.

Before making good this statement, it will be necessary to discuss the question of the dates of 'Hero and Leander' and 'Venus and Adonis'. There is no difficulty in explaining the resemblances between 'Lucrece' and Marlowe's poem. But as regards 'Venus and Adonis' an attempt has been made to place it chronologically before 'Hero and Leander'. In other words, Marlowe is said to have imitated Shakespeare.[1] Is this so?

'Venus and Adonis' was first printed in 1593, having been entered on the Stationers' Registers on the 18th of April of the same year. Hardly six weeks later Marlowe met with a violent death at Deptford, whither he had fled to escape arrest. On 18. May a warrant had been issued against him, while similar proceedings had been taken against associates of his about two months before, as also against the unfortunate Kyd on the 12th of May.[2]—Surely an unfavourable constellation for the composition of such an exquisite, sweet, love-poem as Hero and Leander. If we supposed Marlowe to have known 'Venus and Adonis', we should be obliged to assume that this poem must have been in existence a considerable time before it was printed, and must have been read by Marlowe in a manuscript copy (which in itself would be nothing unusual). An argument to the effect that 'Hero and Leander' must be later than 'Venus and Adonis', for the reason that the latter was in print some time before Marlowe's death, rests on very slippery ground. If Shakespeare's poem was entered on the Stationers' Registers on the 18th of April, it need not necessarily have appeared before the 1st of June, the date of Marlowe's death.

The external reasons are therefore anything but unfavourable to the supposition that Marlowe's composition was known to the poet of 'Venus and Adonis',—a supposition which there are further considerations to strengthen. Is it probable that Marlowe should have imitated his disciple? If he had, how should we explain Shakespeare's great admiration for a poem he had inspired himself? No, Marlowe

[1] Compare Anglia, XXII, 449; and Shakespeare Jahrbuch XVI, 149, and XIX, 249.

[2] Compare Boas's Edition of Kyd, pp. LXVIff., and LXXII; and Dict. of Nat. Biogr., s. v. Marlowe.

is Shakespeare's master alike in dramatic and undramatic poetry. And, judging with an unbiassed mind, we find that 'Hero and Leander' as far surpasses Shakespeare's first heir of his invention, as 'Edward II'. is, in some respects, superior to 'Richard II'.

Marlowe's poem, consisting of 818 verses, was continued by George Chapman, a man of great industry but with most weak hams. It was first printed in 1598. As we find recollections of it in Shakespeare's works before this date, our poet must have had a manuscript copy of it.[1]

Marlowe gives a psychological sketch of the beginning and development of love. Hero and Leander are enamoured of each other at first sight. The latter endeavours to persuade Hero by means of long tirades to give up her virginity. She cries and blushes. Leander continuing his oratory gains her half-unwilling assent, and spends the first night with her in innocent dalliance; not so the second. Say no more. Thus ends Marlowe's 'Hero and Leander', which is in reality a complete little poem of its own.

The reader will not fail to be struck with the general resemblance between Marlowe's sensuous poem and 'Venus and Adonis'. No doubt, 'Hero and Leander' was Shakespeare's model not only in the conception but in sweetness of melody and liquidity of diction and formal excellence generally. At the same time we should not overlook the differences between the two poems. While Shakespeare's theme is love's labour lost, Marlowe's is love's labour won. Shakespeare's poem is characterized by unimpassioned objectivity, which we do not find, nor do we expect to find, in the poet of the *Sturm und Drang*, whose productions are more spontaneous ebullitions of his own inward feelings.

The following are some echoes of Marlowe's poem in Shakespeare's works.

First, I note some general allusions to the story of Hero and Leander. In The Two Gentlemen, I, i, 19, we read:

> *Valentine.* And on a love-book pray for my success?
> *Proteus.* Upon some book I love I'll pray for thee.
> *Val.* That's on some shallow *story of deep love:*
> *How young Leander cross'd the Hellespont.*

---

[1] The practice of circulating MSS. was very common. Compare S. Lee, Life of Shakespeare, p. 88 note.—Marlowe wrote only the first two "sestiads", or cantos, of Hero and Leander. Malone had stated (Var. Ed., XIV, 365), that Marlowe wrote part of the 3rd sestiad. But he withdrew this in vol. XX, 86.

Prot: That's a deep story of a deeper love;
For he was more than over shoes in love.
Val. 'Tis true; for you are over boots in love,
And yet you never *swum the Hellespont.*

Again, Act III, I, 117—120:

Why then, a ladder quaintly made of cords,
.... Would serve to *scale another Hero's tower,*
So *bold Leander* would adventure it.

Compare Hero & L., II, 16:

As he had hope to *scale* the beauteous fort
Wherein the liberal Graces locked their wealth;
And therefore to her tower he got by stealth.
Widee open stood the door; he need not climb.

In Mids. N. Dream, V, I, 198, Pyramus is made to say:

like *Limander* am I trusty still!

to which Thisbe answers in the same strain:

And I like Helen, till the Fates me kill.

Limander is, of course, put for Leander, and Helen for Hero. In 'Romeo', II, IV, 44, we have a humorous allusion to Hero; and, in As You Like It, IV, I, 100—6, Rosalind says jocularly:

Leander, he would have lived many a fair year, though Hero had turned nun, if it had not been for a hot midsummer night; for, good youth, he went but forth to wash him in the Hellespont and being taken with the cramp was drowned: and the foolish chroniclers of that age found it was 'Hero of Sestos'.

Again, in Much Ado, V, II, 30, Benedick mentions

Leander the good swimmer.

The following parallelisms and reminiscences prove how intimately Shakespeare was acquainted with the dainty love-poem.

1) Hero & L., I, 5—8:

Hero the fair,
Whom young Apollo courted for her hair,
And offer'd as a dower *his burning throne,*
Where she should sit, for men to gaze upon.

This expression, as the Rev. Mr. Ebsworth points out,[1] recurs in Shakespeare's Measure for Measure, V, I, 295:

let the devil
Be sometime honour'd for *his burning throne!*

---

[1] Ballad Society, Roxb. Ball., 6, p. 556.

2) The subject of his 'Venus and Adonis' may have been suggested to him by the following lines in 'Hero & L.' (vv. 11—14):

Her wide sleeves green, and border'd with a grove,
Where Venus in her naked glory strove
To please the careless and disdainful eyes
Of proud Adonis, that before her lies.

3) Many would praise the sweet smell as she past,
When 'twas the odour which her breath forth cast.
(Hero & L., I, vv. 21—22).

Compare 'Venus', 443:

For from the stillitory of thy face excelling
Comes breath perfumed that breedeth love by smelling—

and Cymbeline, II, II, 18:

Tis her breathing that
Perfumes the chamber thus.

4) With Hero & L., I, 27:

She ware no gloves; for *neither sun nor wind*
Would burn or parch her hands, but, to her mind,
Or warm or cool them, for they took delight
To play upon those hands, they were so white—

Compare 'Venus', 1081—1092:

Bonnet nor veil henceforth no creature wear!
Nor sun nor wind will ever strive to kiss you:
Having no fair to lose, you need not fear;
The sun doth scorn you and the wind doth hiss you:
But when Adonis lived, sun and sharp air
Lurk'd like two thieves, to rob him of his fair, etc.

5) The expression

Rose-cheek'd Adonis

('Venus', 3) Shakespeare found in 'Hero & L.' (I, 93).

6) Marlowe, describing the fêtes at Sestos (Hero & L., I, 97 ff.), says:

For every street, like to a firmament,
Glister'd with *breathing stars,* who, where they went,
Frighted the melancholy earth, which deem'd
Eternal heaven to burn . . . .
But, far above the loveliest, Hero shin'd
And stole away th'enchanted gazer's mind.

Compare 'Romeo', I, II, 20 ff., where Capulet, adressing Paris, says:—

This night I hold an old accustom'd feast,
Whereto I have invited many a guest . . .
At my poor house look to behold this night
Earth-treading stars that make dark heaven light.
. . . . . hear all, all see,
And like her most whose merit most shall be.

7) Hero & L., I, 158,

There Hero . . .
Vail'd to the ground, veiling her eyelids close—

in conjunction with Hero & L., I, 295:

. . . . . as she spake,
Forth from those two tralucent cisterns brake
A stream of liquid pearl, which down her face
Made milk-white paths, whereon the gods might trace
To Jove's high court—

should be compared with 'Venus', 955ff.:

Here overcome, as one full of despair,
She vail'd her eyelids, who, like sluices, stopt
The crystal tide that from her two cheeks fair,
In the sweet channel of her bosom dropt;
But through the flood-gates breaks the silver rain,
And with his strong course opens them again.

Compare, too, A Lover's Complaint, 281ff.:

This said, his watery eyes he did dismount,
Whose sights till then were levell'd on my face;
Each cheek a river running from a fount
With brinish current downward flow'd apace:
O, how the channel to the stream gave grace!

8) Hero & L., I, 209—214:

This sacrifice, whose sweet perfume descending
From Venus' altar, to your footsteps bending,
Doth testify that you exceed her far,
To whom you offer, and whose nun you are.
Why should you worship her? her you surpass
As much as sparkling diamonds flaring glass.

Compare 'Romeo', II, II, 4ff.:—

. . . . . the envious moon,
Who is already sick and pale with grief,
That thou her maid art far more fair than she;
Be not her maid, since she is envious;

Her vestal livery is but sick and green,
And none but fools do wear it; cast it off.

9) Leander tries to persuade Hero by rational arguments not to consume herself in single life. This were unnatural. The woman is like metal worthless till it is stamped. Beauty must be propagated. These procreation-arguments struck the fancy of the young Shakespeare. Not only are these ideas the theme of many sonnets, but we find their traces in others of his works. Sidney's Arcadia (Bk. III),[1] it is true, contains some passages of a like tendency; but Shakespeare's ideas approach much more nearly to those expressed by Marlowe. By way of example, I quote the following corresponding passages:

Hero & L., I, 317 ff.:

Abandon fruitless cold virginity,
The gentle queen of Love's sole enemy.
Then shall you most resemble Venus' nun,
When Venus' sweet rites are performed and done.
. . . .
Nor stain thy youthful years with avarice.
Fair fools delight to be accounted nice.
The richest corn dies, if it be not reapt;
Beauty alone is lost, too warily kept.

Compare 'Venus', 130:

Beauty within itself should not be wasted;
Fair flowers that are not gather'd in their prime
Rot and consume themselves in little time.

Compare the procreation-sonnets; 'Venus', 163 ff., 751 ff.; All's Well, I, I, 121 *seq.*; Romeo, I, I, 214 ff.; Twelfth Night, I, v, 259. (Comp. Shakesp. Jahrb., XVII, 178; XIX, 184).

With Hero & L. I 265:

Base bullion for the stamp's sake we allow:
Even so for men's impression do we you—

compare The Shrew, IV, IV, 92:

take you assurance of her, 'cum privilegio at imprimendum solum'—

and Meas. for Meas., II, IV, 45:

. . . to remit
Their saucy sweetness that do coin heaven's image
In stamps that are forbid.[2]

---

[1] See Shakesp. Jahrb. XVI, 146. Perhaps more important than the passage in Sidney is one in W. Warner's Albion's England 1589, 5th Book, ch. 24 (end). But Shakespeare does not appear to have consulted Warner either.

[2] Compare also 'Edward III', Act II, I, 262.

10) Hero & L., I, 346—8:

> where all is *whist* and still,
> Save that the sea, playing on *yellow sand*,
> Sends forth a rattling murmur to the land.

This, as Mr. Gosse points out (in The Academy, 5. Dec. 1874), reminds us of Ariel's song in The Tempest, I, II, 376:

> Come unto these *yellow sands*,
> And then take hands:
> Courtsied when you have and kiss'd
> The wild waves *whist*, etc.

11) Hero, we are told (Hero & L., I, 366), had

> hands so pure, so innocent, nay, such
> As might have made Heaven stoop to have a touch.

The same comparison we find in 'Lucrece', 1371—2:

> Ilion . . . Which the conceited painter drew so proud,
> As heaven, it seem'd, to kiss the turrets bow'd.

12) Venus's prophecy (vv. 1135ff.), that sorrow shall attend love, etc., may have owed its insertion to Marlowe's prophecy, at the end of the first 'sestiad', that Learning and Poverty shall go together.

13) Hero & L., II, 1ff.:

> By this, sad Hero, with love unacquainted,
> Viewing Leander's face, fell down and fainted.
> *He kiss'd her, and breath'd life into her lips.*

Compare the effect of Adonis's breath on Venus (473—4):

> For on the grass she lies as she were slain,
> *Till his breath breathed life in her again.*

Again, 'Romeo', V, I, 6:

> I dreamt my lady came and found me dead—
> Strange dream, that gives a dead man leave to think!—
> *And breathed such life with kisses in my lips,*
> That I revived, and was an emperor.

14) Hero & L., II, 141:

> For as a hot proud horse highly disdains
> To have his head controlled, but breaks the reins,
> Spits forth the ringled bit, and with his hoves,
> Checks the submissive ground; so he that loves,
> The more he is restrain'd, the worse he fares.

Compare 'Venus', 259 *seq.*,—the episode of the mare and the stallion, and Venus's application, 'Venus', 385ff., from which I quote vv. 391—6:

How like a jade he stood, tied to the tree,
Servilely master'd with a leathern rein!
But when he saw his love, his youth's fair fee,
He held such petty bondage in disdain;
Throwing the base thong from his bending crest,
Enfranchising his mouth, his back, his breast.

15) Hero & L., II, 171ff.:—

Neptune and the waves desire to rob kisses from Leander swimming in the Hellespont:

the bold waves . . . mounted up, intending to have kiss'd him,
And fell in drops like tears because they miss'd him.

. . . .

He [Neptune] watched his arms, and, as they open'd wide
At every stroke, betwixt them would he slide,
And steal a kiss, and then run out and dance, etc.

Compare 'Lucrece', 386:—

Her lily hand, her rosy cheek lies under,
Cozening the pillow of a lawful kiss;
Who, therefore angry, seems to part in sunder,
Swelling on either side to want his bliss.

And 'Venus', 871:

And as she runs, the bushes in the way
Some catch her by the neck, some kiss her face,
Some twine about her thigh to make her stay:
She wildly breaketh from their strict embrace.

Comp. also 'Venus', 722, 630, & 725, 1080—92.[1]

16) Hero & L., II, 225:

'Tis wisdom to give much; a gift prevails
When deep persuading oratory fails.

Compare The Two Gentlemen, III, i, 89:—

Win her with gifts, if she respect not words:
Dumb jewels often in their silent kind
More than quick words do move a woman's mind.[2]

---

[1] Comp. Jusserand, The Engl. Novel in the Time of Shakespeare, a book of much esprit but without depth, p. 257.

[2] Comp., too, Daniel's Compl. of Rosamond, 376:—

costly Iewels (Orators of Love,)
Which (ah, too well men know) doe women move.

17) Hero & L. II, 269 ff. Leander wants to embrace Hero,

Yet ever, as he greedily assay'd
To touch those dainties, she the harpy play'd,
And every limb did, as a soldier stout,
Defend the fort, and keep the foeman out;
For though the rising ivory mount he scal'd,
Which is with azure circling lines empal'd,
Much like a globe (a globe may I term this,
By which Love sails to regions full of bliss),
Yet there with Sisyphus he toil'd in vain,
Till gentle parley did the truce obtain.
Even as a bird, which in our hands we wring,
Forth plungeth, and oft flutters with her wing,
She trembling strove, etc.

Compare 'Lucrece', 407—473, from which I quote only the following verses:—

(v. 407) Her breasts, like ivory globes circled with blue,
A pair of maiden worlds unconquered.

(v. 437) His hand . . . . . march'd on to make his stand
On her bare breast, the heart of all her land;
Whose ranks of blue veins, as his hand did scale,
Left their round turrets destitute and pale.

(v. 457) Like to a new-kill'd bird she trembling lies.

(v. 481) Under that colour am I come to scale
Thy never-conquer'd fort.

Comp., too, 'Venus', 560.

18) Hero is seemingly coy (Hero & L., II, 283);

Treason was in her thought,
And cunningly to yield herself she sought.
Seeming not won, yet won she was at length:
In such wars women use but half their strength.

Again, Sestiad I, 331:

Hero's looks yielded, but her words made war:
Women are won when they begin to jar.
. . . . . Yet, evilly feigning anger, strove she still,
And would be thought to grant against her will.

This *may* have helped to suggest the following lines respecting the attitude of the mare (Venus, 309):—

Being proud, as females are, to see him woo her,
She puts on outward strangeness, seems unkind;
Spurns at his love and scorns the heat he feels,
Beating his kind embracements with her heels.

19) Hero and Leander, II, 287—290:

Leander now, like Theban *Hercules*,
Enter'd the orchard of th' *Hesperides*;
Whose fruit none rightly can describe, but he
That pulls or shakes it from the golden tree.

Compare for the rhyme *Hercules: Hesperides* Love's Lab. Lost, IV, III, 340:

For valour, is not Love a *Hercules*,
Still climbing trees in the *Hesperides*?

I do not know whether it be a matter of accident or not, that two lines in Greene's 'Friar Bacon and Friar Bungay' (sc. IX, 82—83) end with the same words, though the rest of the entire scene is written in blank verse:

the garden call'd *Hesperides*,
Subdu'd and won by conquering *Hercules*.

20) Hero & L., II, 301:

And now she wish'd this night were never done,
And sigh'd to think upon th' approaching sun;
For much it griev'd her, that the bright day-light
Should know the pleasure of this blessed night.

Compare 'Lucrece', 799ff.:

O Night, thou furnace of foul-reeking smoke,
Let not the jealous Day behold that face
Which underneath thy black all-hiding cloak
Immodestly lies martyr'd with disgrace! . . . .
. . . Make me not object to the tell-tale Day, etc.

21) Hero & L., II, 317 seq.

Thus near the bed she blushing stood upright,
And *from her countenance* behold ye might
A kind of twilight break, which through the air,
As from an orient cloud, glimps'd here and there;
And round about the chamber *this false morn*
Brought forth the day before the day was born.
So Hero's ruddy cheek Hero betray'd.

Compare 'Measure for Measure', IV, I, 1—4:

Take, O, take those lips away,
That so sweetly were forsworn;
And those eyes, the break of day,
Lights that do mislead the morn.[1]

And Romeo, III, II, 8:

Lovers can see to do their amorous rites
By their own beauties; or, if love be blind,
It best agrees with night.

With the latter idea compare Hero & L., I, 191:

dark night is Cupid's day.

22) Finally, the following passage from *Chapman's* continuation, Sest., III, 38:

.... a beauty richer than the sky,
Through whose white skin, softer than soundest sleep,
With damask eyes the ruby blood doth peep,
And runs in branches through her azure veins.

is compared by Malone (XIV, 356) with Wint. Tale, IV, IV, 147:

.... your youth,
And the true *blood* which *peepeth* fairly through't.

This, I think, is the only coincidence worth noting between Chapman's portion and Shakespeare.[2]

[1] Comp. Shakespeariana, Philadelphia, vol. III, 1886, p. 319.

[2] Whether Dyer, the author of Folk Lore of Shakespeare is correct in saying (p. 174) that dragons are a mythological attribute of Night, or whether there is any likelihood of Shakespeare having obtained a suggestion from Marlowe, I cannot decide. The passages in question are these:

Nor that night-wandering, pale, and watery star
(When yawning *dragons* draw her thirling car).
Hero & L., I, 107.

Night's swift dragons cut the clouds full fast.
(Mid. N. Dr., III, II, 379.)

Swift, swift, you dragons of the night.
(Cymbeline, II, II, 48.)

The dragon wing of night o'erspreads the earth.
(Troil. V, VIII, 17.)

According to the old mythology Demeter, or Ceres, alone had a dragon yoke (see Furness, X, p. 165).

THOMAS WATSON.

The passage on Time in 'Lucrece' is inspired by Thomas Watson's 77th poem of his *Hekatompathia*, printed 1582.[1]

I may here add a short note on

SHAKESPEARE'S SONNETS.

I cannot enter into details regarding their sources. This is a question which has received careful attention at the hands of Sidney Lee, Hermann Isaac, and others.[2] The conclusion Mr. Lee arrives at is as follows: 'The thoughts and words of the sonnets of Daniel, 'Drayton, Watson, Barnabe Barnes, Constable, and Sidney were assimi-'lated by Shakespeare in his poems [i. e. sonnets] as consciously and 'with as little compunction as the plays and novels of contemporaries 'in his dramatic work. To Drayton he was especially indebted. Such 'resemblances as are visible between Shakespeare's sonnets and those of Petrarch or Desportes seem due to his study of the English imi-'tators of those sonnetteers. Most of Ronsard's nine hundred sonnets 'and many of his numerous odes were accessible to Shakespeare in 'English adaptations, but there are a few signs that Shakespeare, had recourse to Ronsard direct.' (S. Lee, Life of W. Shakesp., 1898, p. 109.)

SIR PHILIP SIDNEY'S

Arcadia, much read and admired by the Elizabethans, 'has been ob-'viously imitated in many instances by our early dramatists'.

'That part of the pastoral where Pyrocles agrees to command the 'Helots, seems to have suggested those scenes of the Two Gentlemen 'of Verona, in which Valentine leagues himself with the outlaws.' This I quote from Dunlop, who endorses Steevens's supposition, repeated dozens of times by others. But anybody who takes the trouble of reading this tiresome romance will be surprised at the extremely slender resemblance. Pyrocles is indeed captain of the Helots, over whom he exercises a civilizing influence, but there the coincidence ends (Arc. Bk. I, ch. 6). Perhaps a little more apposite would have been a reference to chapt. 5, where the wicked Demagoras is banished, and made the general by the Helots. But I believe Shakespeare never thought of the Arcadia when he wrote Acts IV & V of The Two Gentle-men. The idea of captaining outlaws was suggested by the Robin Hood ballads (s. *post*, Ballads).

[1] Reprinted by Arber.

[2] See also Sarrazin, Shakespeare's Lehrjahre, 1897, pp. 149ff.

'An episode in the second book (ch. 10) of the Arcadia, where a 'king of Paphlagonia, whose eyes had been put out by a bastard son, 'is described as led by his rightful heir, whom he had cruelly used 'for the sake of his wicked brother, has furnished Shakespeare with 'the underplot concerning Gloster and his two sons, in King Lear. 'There are in the romance the same description of a bitter storm, 'and the same request of the father, that he might be led to the 'summit of a cliff, which occur in that pathetic tragedy.' (Dunlop, Hist. of Fict., 1896, II. p. 401—2).

'Steevens pointed out that there are in the first two acts [of 'Pericles] several imitations of ideas in the Arcadia, viz., I, I, 10—11; '62—63; II, I, 63—65; II, 54—55, and last words of scene. The 'passages in the *Arcadia* will be found in the Variorum edition at 'these references. Steevens's further supposition that the name of 'Sidney's hero 'Pyrocles' was the original of our 'Pericles' seems very 'likely.' (P. Z. Round, in Facs. Ed. of Pericles, Quarto I, 1886, p. XII). But Shakespeare is *said not* to have written the first two acts of this play.

The name Mopsa in Wint. Tale may have been suggested by the Arcadia.[1]

JOHN LYLY'S EUPHUES.

Let me begin by quoting the following passage from 'John Lyly and Euphuism'[2] by C. G. Child:—'The discussion of this point leads 'naturally to the question which really originated it—did Lyly and 'Euphuism affect Shakspeare, and did Shakspeare parody Lyly? The 'question, it seems to us, is simple to answer. Euphuism, as we 'took care to note, is a matter of diction, of form, of style, and no-'where in Shakspeare do we find a Euphuistic diction, save in the 'single instance where Euphuism appears to be parodied. In brief, it 'is possible that Euphuism may have exercised some formative in-'fluence upon Shakspeare in his youth, but it at least gave no distinct-'ive quality to his style. A quantitative analysis would be simply 'impossible .... We must agree with Landmann[3] that Shakspeare

[1] For parallelisms between Shakespeare and the Arcadia, cp. Eliza M. West: 'Shaksperian Parallelisms chiefly illustrative of The Tempest and a Mids. N. Dream, collected from Sir Philip Sydney's Arcadia,' privately printed, 1865. Only ten copies were preserved, which was too many.

[2] Münchener Beitr., VII, 1894. It is a pity that Child could not make use of Wurth's work and the paper in the Quarterly Rev. referred to *post*.

[3] see New Shaksp. Soc., Transact. 1880—5, Part II.

'did not parody Euphues in "Love's Labour's Lost", and that the only 'passage in which he did so was "1. Henry IV.", [Act II, IV, 438—461. 'Cp., too, ll. 501—4]. His intention there is clearly proved by his use 'of the famous chamomile metaphor which seems to spread every-'where.' (Child, *ut sup.* pp. 112—'3.)

Lyly's passage containing the camomile metaphor is this: 'Too 'much studie doth intoxicate their braines, for (say they) although 'yron the more it is used the brighter it is, yet silver with much '*wearing* doth *wast* to nothing: though the Cammocke the more it is 'bowed the better it serveth, yet the bow the more it is bent and 'occupied, the weaker it waxeth: *though the Camomill the more it is 'troden and pressed downe, the more it spreadeth*, yet the Violet the 'oftner it is handeled and touched, the sooner it withereth and de-cayeth.' (Arber's Repr., p. 46.)

Child's work was soon after succeeded by Wurth's 'Das Wortspiel bei Shakspere' (Wiener Beitr., 1895) and by a paper contributed to the Quarterly Review, No. 365, 1896, pp. 110ff. by R. Warwick Bond who discovers a great deal more of Euphuism in Shakespeare's works than either Landmann or Child is willing to allow. After referring to the parody of Lyly's style in 1. Henry IV., he continues: 'But it 'seems to have escaped notice that Shakespeare's imitation of the 'style [of Euphues] is by no means confined to this brief parody. 'Other passages exhibit the alliteration, the resemblance of sound, the 'repetition of the same word to give point, above all the antithetic 'or parallel structure of clauses—all the chief marks, in fact, of the 'style except the similes .... It appears first in 'The Merchant of 'Venice', in the talk between Portia and Nerissa and one or two 'other passages, and is constantly reappearing in the work of the next 'few years. It crops up only in the mouths of people of rank and 'education, and chiefly in characters remarkable for wit, such as Fal-'staff, Prince Hal, Portia, Rosalind, Touchstone, or the Clown in ''Twelfth Night'.' .... (Q. R. p. 119).[1]

I have thought it advisable to let these two authorities state their own opinion regarding the influence of style—a question in which the final arbiter is Feeling not Reason, however helpful analyt-

[1] In a note below the text the writer gives numerous examples, from which I select the following as samples: Merch., I, I, 114—'8; II, 1—36; 92—96; 140—'5. 1. Henry IV., Act I, II, 1—5; 26—43; 140—'8; As Y. L. It, I, II, 40—60; 92—6; III, II, 11—32; 46—9; etc.—Mr. Bond repeats his view, as given above, in his edition of John Lyly. 1902. (Cf. vol. I, p. 153.)

ic and critical judgment may be. Ward's statement seems to me very just: 'as to the special characteristics of the Euphuistic style, 'he [Shakespeare] was alike too catholic in his appreciation and too 'eclectic in his appropriation of exotic excellence to imitate Lyly other-'wise than incidentally, or (so to speak) as it might suit himself.' (Engl. Dram. Lit., 1899, I, 281.)

'The Euphuism of Lyly's *plays*', says Mr. Child (p. 88), 'is a simpli-'fied Euphuism'. We must be careful not to confound Lylyan peculiarities and devices with Euphuism. Whether we agree with Child, or not, that Shakespeare's style is not Euphuistic, there can be little question that much of it is Lylyan. The combats of wit, the thrusting and parrying of fantastic conceits, puns and antitheses, as irritating in some plays, as delightful in the more finished comedies, continually remind us of the Lylyan dialogues. But these gymnastics of wit were indulged in by the Elizabethan society at large, and Shakespeare himself is said to have practised wit-combats at the Mermaid Tavern.

Rushton has written a book[1] with a view to prove the great influence of Lyly's Euphues on Shakespeare in thought, language, and phrase. But in many cases the resemblances are 'too slender to warrant any definite conclusion'. The following are some more striking PARALLELISMS:—

I) 'Euphues to Botonio, to take his exile patiently' (Arber's Repr., pp. 186ff.) contains counsels very similar to those given by Gaunt to the banished Bolingbroke, in Richard II. (Act I, III). Read the passages in Lyly and Shakespeare and compare especially the following parallelisms.

> Plato would never accompt him banished *th*at had *th*e Sun, Fire, Aire, Water and Earth, that he had before, where he felt the Winters blast and the Summers blaze, where *the* same Sun, and the same Moone shined, whereby he noted that every place was a country to a wise man, and al parts a pallace to a quiet mind.[2]

Compare with this Rich. II., *ut sup.* ll 275—6:—

> All places that the eye of heaven visits
> Are to a wise man ports and happy havens

[1] 'Shakespeare's Euphuism', 1871. I should have preferred to entitle it: 'Passages from Lyly's Euphues illustrative and elucidatory of Sh'. Rushton's parallelisms ought to be compared with those given in the Quarterly Rev, No. 365 (1896), pp. 125f., and Hense's diffuse articles in Shakesp. Jahrb., VII & VIII. But there is too much of the 'perhaps' and 'possible' in many of these parallelisms.

[2] There is less resemblance in the Friar's advice to Romeus, Brooke, ll. 1443—1467.

And ll. 144—5:

> *Bolingbroke*: .... this must my comfort be,
> That sun that warms you here shall shine on me.

The same thought seems to be echoed in Cymbeline, III, IV 139:—

> Hath Britain all the sun that shines? Day, night,
> Are they not but in Britain? etc.

Again, Euphues says:

> He is to be laughed at which thincketh *the* Moone better at Athens then at Corinth. .... when it was cast in Diogenes teeth, *that* the Sinopenetes had banished him Pontus, yea said he, *I them of Diogenes.*

Comp. Rich. II, ll. 279—'80:

> Think not the king did banish thee,
> But thou the king.

The same thought is in Coriol., III, III, where Coriolanus says to the citizens, who had banished him: "I banish you."

II) There is considerable similarity between the advice given by Euphues to Philautus on board the ship sailing for England (Arber's Repr., p. 246.) and Polonius's advice to Laertes in 'Hamlet' (I, III), the sequence of the counsels paralleling each other being the same in both cases.

Euphues.	*Polonius.*
Be not lavish of thy tongue	Give thy thoughts no tongue.
every one that shaketh thee by the hand, is not ioyned to thee in heart.	But do not dull thy palm with entertainment Of each new-hatch'd, unfledged comrade.
Be not quarrellous for every lyght occasion ... they [i. e. the English] never fight without provoking, and once provoked they never cease. Beware etc.	Beware Of entrance to a quarrel, but being in, Bear't that the opposed may beware of thee.
It shal be there better to heare what they say, then to speak what thou thinkest: They have long ears and short tongues, quicke to heare, and slow to utter.	Give every man thy ear, but few thy voice; Take each man's censure, but reserve thy judgement.

III) Malone pointed out a resemblance between Shakespeare's description of the Commonwealth of the bees (Henry V. Act I, II, 183 ff.) and that given by Lyly in his Euphues (Arber, pp. 261—'5).

Our poet, however, had no doubt learned this object-lesson in the school of Nature.

Further coincidences are pointed out by Mr. Bond in his edition of Lyly (vol. I, p. 164—175), to which I refer the reader.

THOMAS LODGE.

Lodge's 'ROSALYNDE. Euphues Golden Legacie: found after his death in his Cell at Silexedra. Bequeathed to Philautus sonnes, noursed up with their father in England' (printed 1590, and again 1592, 1598, etc.), a novel 'Euphuistic in form and Arcadian in content', is the source of Shakespeare's sweet comedy, As You Like It.[1]

The following poem in Lodge's Rosalynde (Furness, VIII, 366):

> A turtle sate upon a leavelesse tree,
> Mourning her absent pheare
> With sad and sorrie cheare, etc.

seems to be echoed by Paulina, in The Wint. Tale, V, III, 132—5:

> I, an old turtle,
> Will wing me to some wither'd bough and there
> My mate, that's never to be found again,
> Lament till I am lost.

ROBERT GREENE

is another Euphuistic writer, well known to Shakespeare, whose Winter's Tale is founded on Greene's 'PANDOSTO. The Triumph of Time', printed in 1588 and again 1607, 1609, 1614, etc. The running title is 'The Historie of Dorastus and Fawnia' (which in the later editions, 1648 onwards, appeared on the title page).[2]

To Greene's '*Menaphon*' (1589, famous now for the allusion to a Pre-Hamlet, in Nash's forewords to the work) the name 'Menaphon', occurring in The Comedy of Errors (Act V, I, 368), seems to be due.[3]

Of course, Shakespeare must have read Greene's '*Groats-worth 'of Wit*, bought with a Million of Repentaunce', 1592, which contains the earliest undisputed reference to him. Henry Chettle, in his 'Kind-

[1] On Lodge's romance in its relation to Shakespeare's play, see Delius, Shakesp. Jahrb., VI, 226f.; Zupitza, *ib.* XXI, 93ff.; [Boswell-]Stone, New Sh. Soc. Trans., 1880—5, Part II, 227f.

[2] On the relation of Pandosto to Wint. Tale, see Delius, Sh. Jahrb., XV, 22ff.

[3] When I wrote this, I forgot that Menaphon is also the name of a character in Marlowe's Tamburlaine.

Harts Dreame', 1592, seems to indicate, that the 'Shake-scene' was offended with Greene's dying invective.[1]

Of

FRANCIS BACON

I have not been able to discover any traces in Shakespeare's works. On the almost famous passage in 'Troilus and Cressida', II, II, 166:

> young men, whom Aristotle thought
> Unfit to hear moral philosophy—

Mr. Lee, in his Life of William Shakespeare, p. 370, has an excellent note, to which I refer the reader. He shows that Shakespeare and Bacon may well have made a quotation from Aristotle in what at first sight appears to be the same erroneous form independently of each other. Bacon's *Advancement of Learning*, where the quotation in question is to be found, was first published in 1605. 'Troilus' did not appear in print until 1609, but it was probably produced in 1602. Unless we assume that the above passage was inserted after 1605, the observation could not possibly have been borrowed from Bacon.[2]

CAMDEN?

The fable of the Belly and the Members in 'Coriolanus', I, I, is derived from Plutarch, whom Shakespeare follows fairly closely. Certain points of resemblance have also been pointed out between Shakespeare's version and the account in Camden's '*Remains concerning Britain*,[3] published in 1605. In both of these writers the belly is likened to a gulf, and the following two passages contain similar thoughts:

Shakesp.: .... where the other instruments
Did see and hear, divise, instruct, walk, feel.

Camden: whereas the eyes beheld, the ears heard, the hands laboured, the feet travelled, the tongue spake, and all parts performed their functions .....

These ideas are not explicitly set forth by Plutarch. But, though it is possible that Shakespeare read and consulted Camden's entertaining book, the resemblances may be only accidental. No doubt the famous apologue was familiar to the great dramatist in some form or other long before Camden's publication appeared.

On Cordelia's answer in King Lear, see *post*.

[1] See New. Sh. Soc., IV, 2, Cent. of Praise, p. 4.

[2] From the Baconians we learn how *not* to reason. This is some good, though a negative one.

[3] See 'Wise Speeches'. Camden places the tale in the mouth of Pope Adrian, an Englishman by birth, of the family of *Breakespeare*!

SAMUEL HARSNETT.

It was in 1585—1586 that William Weston and other Romanists performed their prodigies of exorcising devils, which, it was hoped, would prove a potent means of converting Protestants The scene of their doings was chiefly in Sir George Peckham's house at Denham, in Bucks. The fiends thus cast out were legion. There could be no mistake about the facts, as there was no lack of eye-witnesses, and the devils had actually been seen swimming like fishes under the skins of the demoniacs. In 1586 Weston, the Jesuit, was imprisoned. Twelve years later there was discovered in the house of a Catholic gentleman 'The Book of Miracles', ascribed to Father Weston, containing an account of the prodigies in question. The book, a manuscript volume,[1] is no longer extant. The chief persons dispossessed by the exorcists were examined or re-examined on their oath by a commission in the year 1602; and the Privy Council ordered Samuel Harsnett to publish an account of the proceedings. This he did in his work entitled: 'A Declaration of egregious Popish Impostures, to 'with-draw the harts of her Maiesties Subiects from their allegeance, 'and from the truth of Christian Religion professed in England, under 'the pretence of Casting out devils. Practised by Edmunds, alias 'Weston a Iesuit, and diverse Romish priests his wicked associates. 'Where-unto are annexed the Copies of the Confessions, and Exami-'nations of the parties themselves, which were pretended to be possess-'ed, and dispossessed, taken upon oath before her Maiesties Com-'missioners, for causes Ecclesiasticall. At London Printed by Iames 'Roberts, dwelling in Barbican. 1603.' The work appeared again in 1604, and in 1605. Regarding Weston, Harsnett, etc., see Mr. T. G. Law's paper on 'Devil-hunting in Elizabethan England', in The Nineteenth Century, March, 1894.[2]

Theobald was the first to point out that Harsnett's book supplied the materials for the diablerie of the scenes in King Lear, in which Edgar assumes the *rôle* of Mad Tom.

In Harsnett's work we find, among many other names of devils, the following: Frateretto, Fliberdigibbet, Haberdicut or Hoberdicut,

---

[1] See Harsnett p. 1: 'the *penned* booke of Miracles' .... 'an English Trea-'tise in a *written* hand.' Harsnett gives numerous quotations from it.

[2] Cf. also Spalding, Elizabethan Demonology.

Hoberdidance, Maho, Modu, Purre, Smolkin. These are all in King Lear. In Act III, iv, 120, we read:—

> *Edgar.* This is the foul fiend *Flibbertigibbet.*

In the modern editions of King Lear we find the name again in Act IV, i, 62, where the Quartos have '*Stiberdigebit*'.—Again, compare Act III, iv, 146 and 149:—

> *Edg.* . . . Peace, *Smulkin*; peace, thou fiend!
> The *prince* of darkness is a gentleman:
> *Modo* he's call'd, and *Mahu.*[1]

And Act III, iv, 7:

> *Edg. Frateretto* calls me; and tells me Nero is an angler in the lake of darkness. Pray, innocent, and beware the foul fiend.

In l. 32 occurs the name *Hoppedance*, for which the modern editors write Hopdance. In l. 47 we have '*Pur*! the cat is gray', which may be an allusion to Harsnett's devil, Purre. In Act IV, i, 60—5, Edgar is made to say:—

> Five fiends have been in poor Tom at once; of lust, as *Obidicut; Hobbididance*, prince of dumbness; *Mahu* of stealing; *Modo*, of murder; *Flibbertigibbet*, of mopping and mowing, who since *possesses chambermaids and waiting-women.*

In this last sentence there is no doubt a reference to the maids of the Peckham family, where the exorcists carried on their tomfoolery. Sara Williams was a special favourite of the devils. When she was 'troubled with a wind in her stomacke, the priests would say 'at such times, that then the spirit began to rise in her.' And 'if 'they heard any *croaking in her belly* . . . then they would make a 'wonderful matter of that.' At one time the priests declared the croaking to be a devil speaking 'with the voyce of a Toade.' (Cp. Harsnett, pp. 194—5.) This Shakespeare had in mind when he made Edgar say (Act III, vi, 31—4):—

> The foul fiend haunts poor Tom *in the voice of* a nightingale. *Hopdance cries in Tom's belly* for two white herring. *Croak not, black angel;* I have no food for thee.

'Ma: Maynie', says Harsnett (p. 25; cp. too p. 263) 'had a spice of 'the *Hysterica passio*, as seems from his youth, hee himselfe termes

[1] According to Harsnett Maho and Modu were the chief devils that possessed Sara and Maynie, cf. pp. 48, 50, and p. 269, where Harsnett mentions 'the great 'Prince Modu'.

'it the Moother.' This suggested the following passage put into the mouth of Lear (Act II, IV, 56–8):

> O, how this *mother* swells up toward my heart!
> *Hysterica passio*, down, thou climbing sorrow,
> Thy element's below!

The same Maynie, one of the demoniacs, who used to curl his hair was affirmed by the exorcist to be troubled with the spirit of Pride. The Jesuit succeeded in casting this spirit as well as seven more out of him, every one of which departed from him in the shape of some animal: 'the spirit of Pride departed in the forme of a Peacock. 'The spirit of Sloth in the likenes of an Asse: the spirit of Envie in 'the similitude of a Dog: the spirit of Gluttony in the forme of a 'Wolfe: and the other devils had also in their departure their parti-'culer likenesses agreeable to their natures'. (Harsnett, p. 281.) Compare with this, Act III, IV, 86:

> *Lear.* What hast thou been?
> *Edg.* A serving-man, *proud* in heart and mind; that *curled my hair*; . . . . . hog in sloth, fox in stealth, wolf in greediness, dog in madness, lion in prey.

The idea, however, of typifying sins by animals is old. Compare the Clarendon Press edition of King Lear, p. 169.

On p. 219, Harsnett relates that a halter and two blades of knives were left by some one in the gallery floor of Peckham's house. Maynie, in his fit, said, 'that the devil layd them in the Gallery, that some 'of those that were possessed, might either hang theselves with the 'halter, or kil themselves with the blades'. This seems to be alluded to in King Lear, III, IV, 54, where Edgar speaks of the foul fiend,

> that hath laid knives under his pillow, and halters in his pew.

The pregnant words 'corky' (III, VII, 28), 'star-blasting' (III, IV, 60), 'mopping and mowing' (IV, I, 63, the Quartos read: 'Mobing and 'Mohing') may also have been suggested by Harsnett, who uses the word 'corky', applied to an old woman, on p. 23, and 'sparrow-'blasting' and 'sprite-blasting of the devil', on p. 80. On p. 136, Harsnett says: If 'she have a little helpe of the *Mother, Epilepsie*, or '*Cramp*, to teach her role her eyes, wrie her mouth, gnash her teeth, 'startle with her body, hold her armes and hands stiffe, make anticke 'faces, girne, *mow, and mop like an Ape, tumble like a Hedgehogge,* 'and can mutter out two or three words of gibridg, as *obus, bobus* . . . . 'then no doubt . . . . . the young girle is *Owle-blasted*, and possessed'.

Douce thinks there is a reminiscence of this passage in The Tempest, II, II, 9, where Caliban speaking of Prospero's spirits says:

> For every trifle are they set upon me;
> Sometime like apes that mow and chatter at me,
> And after bite me; then like hedgehogs which
> Lie tumbling in my barefoot way . . . . .

(Comp. also Tempest, IV, I, 47.) Probably the coincidence pointed out by Douce is accidental.

FURTHER DÆMONOLOGICA.

Having pointed out Shakespeare's connexions with Harsnett, we may here take note of further demonological matter in his works. Many dramas of the time, we know, derive themes and hints from the supernatural and demonological world of thought,—Marlowe's 'Dr. Faustus' deserving especial mention.[1] We need therefore be little surprised to find Shakespeare introducing *motifs* of this sort into his plays.

Of 'Lear' I have already spoken. In 'Macbeth' and 'Tempest' the supernatural machinery forms an essential element of the play. In 1. Henry VI., Act V, III, the Maid of Orleans invokes her familiars; but they forsake her even though she offers all:

> Then take my soul, my body, soul and all.

No doubt this is an echo from Marlowe's 'Dr. Faustus', whose influence is still more apparent in 2. Henry VI., Act I, IV, from which I quote the following passage: ll. 24—34:

> *Bolingbroke.* Madam, sit you and fear not: whom we raise,
> We will make fast within a hallow'd verge.
> [*Here they do the ceremonies belonging, and make the circle; Bolingbroke or Southwell reads,* Conjure te, &c. *It thunders and lightens terribly; then the Spirit riseth.*]
> *Spirit.* Adsum.
> *M. Jourdain.* Asmath,
> By the eternal God, whose name and power
> Thou tremblest at, answer that I shall ask;
> For, till thou speak, thou shalt not pass from hence;
> *Spir.* Ask what thou wilt. That I had said and done!
> *Boling.* 'First of the king: what shall of him become?'

---

[1] Compare my remarks in the Archiv für das Studium der neueren Sprachen, vol. CVII, p. 182.

Spir. The duke yet lives that Henry shall depose;
But him outlive, and die a violent death, etc.

In 'Comedy of Errors' (IV, IV), 'Twelfth Night' (III, IV and IV, II) and 'Romeo' (II, I) we have humorous conjurations and exorcisms 'A Midsummer Night's Dream' takes us to the serene fairy-land, into which we also peer through a chink, so to speak, in 'Romeo' (I, IV) and 'Comedy of Errors' (II, II, 190—204). In The Merry Wives Falstaff is pinched black and blue by supposed fairies, like Corsites in Lyly's Endymion. From the undiscovered country we see travellers return as revengers in 'Richard III.', 'Hamlet', 'Julius Caesar' and Macbeth.[1] And, lastly, we have the apparitions in Cymbeline (V, IV) and Henry VIII. (IV, II).

Devils were classified into greater and lesser ones. To the former class belong, *inter alios*, Lucifer, Beelzebub, Amaymon king of the East, Zimimar king of the North[2], and apparently also Barbason [=Barbas?] and Asmath [=Asmodeus?]. These are all mentioned or referred to by Shakespeare.[3] To the lower rank belong Paddock, Graymalkin, Harpier, etc.

Oberon, Titania, and Puck *alias* Hobgoblin *alias* Robin Goodfellow are fairies. Ariel is a spirit, almost a fairy. Caliban is the offspring of a hag and an incubus.[4]

[1] The ghost comes from Seneca. Compare above, p. 36.

[2] Compare Reg. Scott, ed. Nicholson, p. 327, 314, 318, 321. Randle Holme (Academy of Armory. 1688. 2. Bk. p. 13) makes Amaymon ruler of the North part of the Infernal gulf and Zimimar of the East part.

[3] Merry Wives, II, II, 311; 1. Henry IV., Act II, IV, 370—1; Henry V., Act II, I, 57; IV, VII, 145; 1. Henry VI., Act V, III, 6 (*the lordly monarch of the north*); 2. Henry VI., Act I, IV, 27; etc.

[4] Hecate in 'Macbeth' is probably not Shakespeare's creation. Nor do the songs 'Come away, etc.' and 'Black spirits, etc.' (Act III, V; and IV, I), given in full in Middleton's Witch, appear to have formed part of the original draught of 'Macbeth'. See Herford, Eversley Edition of Shakespeare, vol. IX. To Herford's arguments I am in position to add another. Nowhere in Shakespeare's Macbeth do we find our poet *closely* following Scott. But in Middleton's 'Witch' there are many expressions taken unaltered from 'The Discovery', and the song 'black spirits' happens to be one for which the unmistakeable source can be pointed out in this work (p. 455). Hoppo, Stadlin, Puckle, and Hellwain of the other 'song' are likewise taken from Scott (See Nicholson's edition, p. 551)—Remember, that Middleton's 'Witch' belonged to "His Maties Servants", the troupe of which Shakespeare was a member. Regarding Hecate compare, too, Herford, 'Literary Relations between Germany and England' (1886) p. 235. Herford's observations on witchcraft, ibid. p. 219 ff., form a valuable complement to T. A. Spalding's work on Elizabethan Demonology. 1880.

Demonological problems occupying people's minds then to so large an extent both in practice and theory, we need not search far for the sources of Shakespeare's witchcraft-knowledge. REGINALD SCOTT's *The Discovery of Witchcraft*, however, printed in 1584, deserves to be noted as an exhaustive work on this subject,—a compendium which, we know, was used by Ben Jonson[1] and Middleton; and which was doubtless also referred to by Shakespeare, though I cannot prove it decisively. Numerous passages in illustration of Shakespeare might be easily cited from Scott's work.[2]

In 'Macbeth' Shakespeare makes large concessions to the idiosyncracies and whims of the new king, James I., one of the foremost champions in the quixotic fight against the black art; and, unless we are mistaken, the play contains reminiscences of historic events of 1589—1591 in which the king played a prominent figure. The facts are these. In the year 1589 Anne of Denmark, King James's bride, sailing for Scotland, was driven upon the coast of Norway by a violent tempest. James thereupon put to sea himself to fetch her home. During their voyage the royal couple experienced another storm. No doubt that was the work of Scotch witches and wizards! By means of threats and tortures the king succeeded in forcing some poor folks to make the most astounding confessions. One Agnes Sampson openly avowed:—

'that at the time when his Majestie was in Denmarke, shee being accompanied by the parties before speciallie named, tooke a cat and christened it, and afterward bounde to each part of that cat, the cheefest parte of a dead man, and severall joyntis of his bodie: And that in the night following, the saide cat was *convayed into the middest of the sea by all these witches, sayling in their riddles or cives* (=sieves), as is aforesaid, and so left the saide cat right before the towne of Lieth (*sic*) in Scotland. This doone, there did arise such a tempest in the sea, as a greater hath not bene seene; which tempest was the cause of the perishing of a boat or vessell comming over from the towne of Brunt Ilande to the towne of Lieth, wherein was sundrie jewelles and rich giftes, which should have been presented to the now Queene of Scotland, at her Majesties coming to Leith. Againe, it is confessed, that the saide christened cat was the cause that the Kinges Majesties shippe, at his comming forth of Denmarke, had a contrarie winde to the rest of his shippes then being in his companie; which thing was most strange and true, as the Kinges

---

[1] Comp. my note in Shakespeare Jahrbuch, XXXVIII, p. 240.

[2] Concerning the charm on p. 68 of Scott's 'Discovery', see *post*.

Majestie acknowledgeth, for when the rest of the shippes had a faire and good winde, then was the winde contrarie and altogether against his Majestie; and further, the sayde witche declared, that his Majestie had never come safely from the sea, if his faith had not prevayled above their intentions.[1]

Who will not be reminded of Shakespeare's 'Macbeth', I, III?

First Witch .... Her husband's to Aleppo gone, master o' the Tiger:
But in a sieve I'll thither sail,
And, like a rat without a tail,
I'll do, I'll do, and I'll do.
Sec. Witch. I'll give thee *a wind*, etc.
. . . .
First Witch .... Though his bark cannot be lost,
Yet it shall be tempest-tost, etc.

'It is worth a note that this art of going to sea in sieves, which 'Shakspere has referred to in his drama, seems to have been pecu-'liar to this set of witches. English witches had the reputation of 'being able to go upon the water in egg-shells and cockle-shells [cp. 'Reg. Scott, p. 8 and 'Pericles', IV, IV, 2], but seem never to have 'detected any peculiar advantages in the sieve.'[2]

An account of the intensely exciting proceedings at Edinburgh was printed in London in a pamphlet entitled: *Newes from Scotland*, declaring the Damnable life and death of Doctor Fian,[3] a notable

[1] I am quoting from 'Newes from Scotland' (see below), republished in R. Pitcairn's Criminal Trials in Scotland, vol. I, part 2, p. 218. Comp. also *ibid.*, p. 254 (25).

[2] Spalding, Elizabethan Demonology, 1880 p. 115. In this book the reader will find further particulars regarding these events and their connexion with 'Macbeth'.

[3] Poor Dr. Fian was a young schoolmaster, who refused in the end to confess to a crime punishable by death. Even under the most frightful tortures, when his finger-nails were wrenched off by a pair of pinchers, and his legs crushed into one mass of bleeding jelly, did he remain 'stubborn'. 'So deeply had the Devill entered into his heart' (Pitcairn, *ut sup.*, p. 223). The only alternative, therefore, that remained was to kill him. More they could not do. Other supposed associates, too, of this martyr-against-his-will received what their hellish deeds deserved.

But King James was not the only devil-hunter. In Queen Elizabeth's reign a considerable number of witches were put to death (see Mrs. Lynn Linton, Witch Stories. 1883, p. 153; and Annie Besant, Threatenings and Slaughter's, London 1886, p. 21.); and in Germany, Heinrich Julius Duke of Braunschweig-Wolfenbüttel, who married the sister of King James's wife, was notorious for his wholesale executions of innocent folks. But let us cast a mantle of pity over the misdeeds

Sorcerer, who was burned at Edenbrough in January last, 1591 .... With the true examinations of the saide Doctor and witches as they uttered them in the presence of the Scottish King. Discovering how they pretended to bewitch and drowne his Maiestie in the Sea comming from Denmarke, with such other wonderful matters as the like hath not been heard of at any time. Published according to the Scottish copie. At London Printed for Thomas Nelson.'—(Another edition was printed for William Wright.)

In 1597 James published his *Dæmonologie*—'a work founded to 'a great extent upon his experiences at the trials of 1590'. The work was reprinted in 1600, and in 1603; and translated into Latin. Old Nick had therefore tough times of it, while James was king.

BOOKS ON GOOD MANNERS AND ON DUELLING.

In As You Like It, V, IV, 95, Touchstone is made to refer to the

books for good manners,

of which there existed many in Shakespeare's days, as, for example, Hugh Rhodes's *Boke of Nurture* (*c.* 1545, etc.), Francis Seager's *Schoole of Vertue* (1557, 1588), and so forth. (See Dr. Furnivall's 'The Babees Book', edited for the Early English Text Society, No. 32—out of print—; and 'Queene Elizabethes Achademy', Extra Series, VIII.)

In 'Romeo' (I, v, 112)—

You kiss by the book—

we have another reference (if the expression is to be taken literally) to books on manners.

Bacon, in a speech of 1613 against duels, refers to "some French "and Italian pamphlets, which handle the doctrine of *Duels*".[1] The best authority in England on this subject seems to have been Vincentio Saviolo,[2] an Italian, who was the author of 'The booke of honour and Armes wherein is discoursed the Causes of quarrell, and

of our forefathers.—Remember, that they based their firm conviction of the existence and efficacy of the black art and of possession by spirits on the Scriptures. Even Reginald Scott finds it hard to explain the tale of the witch of Endor (1. Sam. 28.) on rational grounds. It is very probable that Shakespeare had this very story in his mind when he wrote Act IV, sc. 1. of Macbeth.

[1] Bacon's Works ed. by Spedding and Ellis and Heath, vol. XI, 1868, p. 400.

[2] Saviolo was taken into the service of the Earl of Essex, and was considered the best fencer in England. Compare what Dekker says in his 'Wonderfull Yeare', 1603 (Huth Libr., Dekker, vol. I, p. 120), of sickness: "Hees the best "Fencer in the world: *Vincentio Saviolo is no body to him*".

the nature of Iniuries with their Repulces with the meanes of satisfaccion and pacification', etc. 1589.[1] This work is not extant. But Malone, it appears, saw it, but he found nothing in it worthy of much note. Another work, appeared in 1594-5 with the title of 'Vincentio Saviolo his Practise. In two Bookes. The first intreating of the use of the Rapier and Dagger. The second, of Honor and honorable Quarrels'. London 1595.

The affectations connected with duelling are derided by Shakespeare in 'As You Like It', V, IV, and 'Romeo', II, IV, 20—37,[2] perhaps with a distinct reference to Saviolo's books. For correspondences between Shakespeare's text and Saviolo I refer the reader to Furness's Variorum Edition.[3]

SHAKESPEARE AND THE EMBLEM WRITERS.

Henry Green's work on this subject (printed in 1870, and comprising 571 pages) gives an incorrect impression of Shakespeare's obligations to the emblematists. Except for Pericles, Act II, II,—but Acts 1 and 2 Shakespeare is said *not* to have written—there is no reliable indication of the poet's indebtedness to the emblem literature. If there is anything to be learnt from Green's volume, it is this, that Shakespeare may have acquired some of his knowledge at second hand.[4]

COMPENDIUMS OF ENGLISH HISTORY.

The problem of the sources of Shakespeare's dramas dealing with English History has been exhaustively investigated by Mr. Boswell-Stone in his 'Shakspere's Holinshed', to which my Inaugural Dissertation (1900) serves as a sort of introduction. To these works I refer the reader for particulars. (Compare also the 'Synopsis' on page 1.) I only state here, very briefly, that Shakespeare has essentially followed RAPHAEL HOLINSHED'S *Chronicles*, in the second edition which appeared in 1586-7. He also made use of EDWARD HALL'S *Union of the Two Noble and Illustre Famelies of Lancastre and York*, probably

[1] Arber Transcripts, II, 537.

[2] Comp., too, Love's L. L., I, II, 184 ff.

[3] In the Dict. of Nat. Biogr. Mr. Lee refers to a note by Ingleby in the Notes and Queries, who imagined he had discovered a source for Orlando's duel with Charles in Saviolo. But Shakespeare had no need to go further than to Lodge's Rosalynd for this episode.

[4] Von Mauntz is the author of Sh. und die Heraldik. Berlin, 1903. (His explanation of The Phoenix and the Turtle seems over-ingenious.)

as embodied in RICHARD GRAFTON'S *Chronicle*, and perhaps also of JOHN STOWE'S *Annales* and FABYAN'S *Chronicle*. FOXE'S *Acts and Monuments* were largely laid under contribution by the author, or authors, of 'Henry VIII'.[1]

[1] An incident from earlier French history (1404) is introduced into Love's Labour's Lost (Compare Hazlitt, Shakesp.'s Libr. I, 3).

Allusions to Contemporary Incidents and Personages, etc. in Shakespeare's Works.— This ought to be made the subject of a separate paper. I only jot down a few notes here from memory. First, in Love's Labour's Lost: Henry of Navarre (Henry IV. of France); Biron; Longaville; Duc du Maine; La Mothe; Duke Alençon; 'Fantastical Monarcho'; a Russian courtship at Queen Elizabeth's court in 1583; and a meeting of Henry of Navarre with a French princess (see Gent. Mag., Oct., 1880, pp. 447—458, and a long article in Sh. Jahrb. XXXI, 200).—Count Mömpelgard (Merry Wiv.); the Armada (Lov. Lab. L.); English and Dutch explorations (see a later chapter); French civil war (Errors, III, II, 125-7); Amurath of Turkey (2. Henry IV.); Gonzaga's murder near Mantua (? see Sh. Jahrb. XXXI, 169); Queen Elizabeth (Henry VIII., Act V, V), "the imperial votaress" at Kenilworth (Mids. N. Dr.), "our gracious empress" (Henry V., Act V, Prol., 30), "the queen" (2. Henry IV. Epil.),—the Queen was probably also before the poet's mind when he drew the Princess in Love's Lab. Lost; King James in Henry VIII., *ut sup.*, Macbeth (especially Act IV, I, 121),— probably also before Shakespeare's mind's eye when portraying the duke in Meas. f. M. and Prospero in The Tempest (cf. Sh. Jahrb. XXXV., 166 and Archiv. CVII, 177); Earl of Southampton (Dedications of 'Venus' and 'Lucrece', Sonnets); Southampton's mother (Mids. N. Dr., cf. Archiv XCV, 291); Earl of Essex and the Irish war (Henry V.); Marriage at court (Tempest, see Shakesp. Jahrb. XXXV *ut sup.*); lady of the Strachy (? Tw. Night); Marlowe, the "dead shepherd" (As Y. L. It); Sneak's noise (2. Henry IV., II, IV, 12); Mistress Mall (Tw. Night—see Dyce's Glossary, ed. 1902); the dark lady of the Sonnets; Brownist (Tw. Night); Puritans (passim); Sackerson, the bear (Wives); Marocco, "the dancing horse" (Love Lab. L.); [Dowland and Spenser occur in a *non*-Shakespearean Sonnet in Pass. Pilgrim 107, 109]; Sir Oliver *Mar*text may contain a distant allusion to the *Mar*-Prelate Controversy. There is a reference to the fierce stage-quarrel, which was being waged about 1600, in Hamlet (II, II). Several actors' names of the Shakespearean troupe are preserved in the Folio of 1623: Kemp, Cowley, Sincklo, Humphrey [Jeffes], Gabriel [Spenser] and Jack Wilson (see Sidney Lee's Facsimile of the Folio, 1902, p. XX). Giulio Romano (Wint. Tale) lived before Shakespeare's time (d. 1546).

CHAPTER 4.

THE ENGLISH DRAMA.

MUCH attention has of late years been paid to Elizabethan dramatic literature. But no comprehensive treatise has yet been written on Shakespeare's relations to it, a subject of utmost importance. Nor can I hope to supply the want of such a *desideratum* in the present chapter, in which I only bring forward what I consider to be some more prominent facts. I make no attempt at tracing out the various currents of the older drama, which meet in Shakespeare. This task belongs rather to the writer of the history of the drama. Much excellent work, I am glad to say, has been done in this direction by men like Ward, Fleay[1], Brandl, Boas, and others, not to speak of critics of bygone days. Suffice it to say here, that Shakespeare, so far from being the creator of a new drama, is the inheritor of an extensive stage-literature, which had attained a considerable degree of perfection. In the present chapter I confine my sole attention to those particular plays of which we find distinct and clear traces in Shakespeare's works, or which are his direct sources.

DRAMATIC AUTHORS.

The three most prominent predecessors of Shakespeare in the English dramatic literature are named by Ben Jonson in his dedicatory verses before the folio of 1623:—

"thou didst our *Lily* out-shine,
"Or sporting *Kid,* or *Marlowes* mighty line."

---

[1] My statement above, p. 5, note, regarding a new edition of Fleay's History of the Stage requires a slight correction. Prof. Morsbach has kindly informed me, that it is an independent work, bearing the following title: Die Geschichte der englischen Schauspiel-Truppen bis 1642. Louvain. 1903. The author is Dr. Maas. It is now passing through the press.

We begin with the last, who stood head and shoulders above his compeers,

CHRISTOPHER MARLOWE,

with whom Shakespeare came into immediate contact, and from whom he received influences the most profound and enduring. Nor was this influence merely *intellectual*, but it is the power and virtue of genius which went out of the poet of the *Sturm und Drang* to his great disciple, gifted with a double portion of his spirit.

Marlowe's strongly individualistic boldly relieved characters have left a marked impression on Shakespeare the character-poet *par excellence.* The aspiring Machiavellian figures, like Tamburlaine, the Duke of Guise, are the prototypes of York in 'Henry VI.', of Richard III., Macbeth, etc. The weak kings, like Mycetes, and Edward, are the models of Shakespeare's Richard II., and Henry VI. Again, Aaron, Shylock, Iago, and others are drawn on the same lines as Barabas, the Jew of Malta, and his accomplice Ithamore. The pair Isabella-Mortimer in Marlowe's Edward II. preceded Margaret-Suffolk and Tamora-Aaron. And, lastly, the murderers of Clarence and of Banquo are modelled after the murderers in Edward II. and in The Massacre at Paris.

The metre chosen by Marlowe was the blank verse, which he was not the creator of, but on which he impressed his peculiar stamp, proving its adaptedness for dramatic purposes.

We now proceed to a brief discussion of Marlowe's works in particular, taking special note of striking parallelisms. Such parallel passages, I admit, do not show the extent of Shakespeare's indebtedness, but they help to bring the fact of it into strong relief. The existence of numerous verbal reminiscences of older plays is just what we might expect to find in a poet, who, as an actor, had taken part in performing many of them dozens of times.[1]

[1] E. Hübener has written a dissertation 'Der Einfluss von Marlowe's Tamburlaine auf die zeitgenössischen und folgenden Dramatiker'. Halle. 1901. Shakespeare's relations to Marlowe are discussed incidentally by Sarrazin in his Shakespeare's Lehrjahre. 1897. Brandl has written some suggestive remarks on the same subject in the Göttingische gelehrte Anzeigen, 1891, p. 712 ff. in a review of Schröer's treatise on Titus Andronicus. Mr. Verity's 'Influence of Marlowe on Shakespeare', 1886 (mentioned in the Dict. of Nat. Biogr.), is a schoolboy's essay.— My quotations from Marlowe are taken from Bullen's edition, 1885.

TAMBURLAINE.

In Shakespeare's earlier plays we are continually reminded of Marlowe's plays. Tamburlaine's influence is predominant in 1. Henry VI., in which we have plenty of fighting and ranting. Many coincidences will be discovered. It will be sufficient to point out the following The impetuous Talbot, the "terror and bloody scourge" of the French, is plainly drawn after Tamburlaine. Suffolk falls in love with his fair prisoner, like Tamburlaine with Zenocrate.[1] The beginning of the first scene of 1. Henry VI. contains echoes from Marlowe's play. Compare especially verses 1—5:

[Dead March. Enter the Funeral of King Henry the Fifth, etc.
Bedford. Hung be the heavens with black, yield day to night!
Comets, importing change of times and states,
Brandish your crystal tresses in the sky,
And with them scourge the bad revolting stars
That have consented unto Henry's death.

The bringing of the coffin on the stage accompained by lamentations is imitated from Marlowe's 2. Tamburlaine, Act III, II, where the hearse of Zenocrate is brought in. The same *motif* occurs in Titus Andronicus (I, I) and Richard III. (I, II). The words placed in the mouth of Bedford appear to contain a reminiscence of 2. Tamburlaine, Act II, IV,—Zenocrate's death-scene:—

Tamb. Black is the beauty of the brightest day;
The golden ball of heaven's eternal fire,
That danced with glory on the silver waves,
Now wants the fuel that inflamed his beams;
And all with faintness, and for foul disgrace,
He binds his temples with a frowning cloud
Ready to darken earth with endless night.
Zenocrate . . . .
Now by the malice of the angry skies,
Whose jealousy admits no second mate,
Draws in the comfort of her latest breath—

or of similar language used in 2. Tamburlaine V, III, 1ff.

The verse

Brandish your crystal tresses in the sky

[1] This *motif* occurs again in Titus Andronicus, I, I, where Saturninus makes Tamora his queen.

is an echo of 1. Tamburlaine, Act V, I, 141:

> Shaking her silver tresses in the air.

The bombastic enumeration of titles and names in Tamburlaine are the object of ridicule in 1. Henry VI., Act IV, VII, 60 *seq.*:—

> But where's the great Alcides of the field,
> Valiant Lord Talbot, Earl of Shrewsbury,
> Created, for his rare success in arms,
> Great Earl of Washford, Waterford and Valence;
> Lord Talbot of Goodrig and Urchinfield, etc., etc.

on which the Pucelle remarks:

> Here is a silly stately style indeed!
> The Turk, that two and fifty kingdoms hath,
> Writes not so tedious a style as this.

In illustration of the above passage we might cite several from Marlowe's play. The following from 2. Tamburl., III, I, 1—7, is perhaps specially characteristic:

> Callapinus Cyricelibes, otherwise Cybelius, son and successive heir to the late mighty emperor, Bajazeth, by the aid of God and his friend Mahomet, emperor of Natolia, Jerusalem, Trebizond, Soria, Amasia, Thracia, Illyria, Carmania, and all the hundred and thirty kingdoms late contributory to his mighty father. Long live Callapinus, Emperor of Turkey.

So much for Henry VI. Part 1, which some critics think was not composed, but only retouched, by Shakespeare. This is a question which we cannot decide with certainty. But there seem to be more hands than one in the play. On the other hand, I am disposed to believe that Shakespeare was the author of Henry VI., Parts 2 and 3, and that 'The First part of the Contention', etc., and 'The True Tragedy of Richard Duke of Yorke' are imperfect reports of the former plays.

The following passage from 3. Henry VI. Act I, II, 28—31

> Therefore, to arms! And, father, do but think
> How sweet a thing it is to wear a crown;
> Within whose circuit is Elysium
> And all that poets feign of bliss and joy.

ought to be compared with 1. Tamburlaine, Act II, v, 51:

> *Tamb.* . . . . Is it not passing brave to be a king,
> "And ride in triumph through Persepolis?"
> *Tech.* O, my lord, 'tis sweet and full of pomp.

Usum. To be a king is half to be a god.
Ther. A god is not so glorious as a king.
I think the pleasure they enjoy in heaven,
Cannot compare with kingly joys in earth.—
To wear a crown enchased with pearl and gold,
Whose virtues carry with it life and death, etc.

The following lines in 2. Henry IV., Act II, IV, 178, placed into the mouth of Pistol:

And hollow pamper'd jades of Asia,
Which cannot go but thirty mile a-day—

contain a burlesque allusion to the opening of the fourth scene of Act IV. of 2. Tamburlaine, in which the hero of the play 'drawn 'in his chariot by the Kings of Trebizond and Soria, with bits in their 'mouths, reins in his left hand, and in his right hand a whip with 'which he scourgeth them', etc., exclaims—

Tamb. Holla, ye pampered jades of Asia!
What! can ye draw but twenty miles a day,
And have so proud a chariot at your heels . . . . .

As further parallelisms I mention 1. Tamburlaine II, II, 72:

Strike up the drum! and march courageously!
Fortune herself doth sit upon our crests.

and Richard III., Act V, III, 351

Upon them! victory sits on our helms!

or Rich. III., Act V, III, 79:

Fortune and victory sit on thy helm.

Again, 2. Tamburlaine V, III, 225:

[They bring in the hearse of Zenocrate.
Tamburlaine (on the point of dying): Now eyes enjoy your latest benefit.

and Romeo, V, III, 112:

Eyes, look your last!
Arms, take your last embrace![1]

The Jew of Malta.

Marlowe's Barabas is the prototype of Shakespeare's Shylock, as has been often remarked. No Barabas, no Shylock![2] The novel of

[1] Further echoes from Tamburlaine in Shakespeare's earlier works are pointed out by Hübener *ut. sup.*

[2] Such as he is. Of course, there was a Jew in the Pre-Merchant.

the Pecorone, on which part of the plot of the Merchant of Venice is based contains no finished delineation of the character of the Jew, and has no daughter, for whom the original will be found in Abigail, the Jew of Malta's daughter.[1] A number of parallelisms between the two plays have been drawn up by A. W. Ward (Hist. of Dram. Lit., I, 346). I refer to the following coincidence by way of example:

Shylock. Signior Antonio, many a time and oft
In the Rialto you have rated me
About my moneys and my usances:
Still have I borne it with a patient shrug,
For sufferance is the badge of all our tribe.
You call me misbeliever, cut-throat dog,
And spit upon my Jewish gaberdine,
And all for use of that which is mine own.
(Merchant, I, III, 107.)

Compare Marlowe's Jew, II, III, 23:—

I learned in Florence, how to kiss my hand,
Heave up my shoulders, when they call me dog.

The following correspondence has also been pointed out. In 'The Jew', III, II, 11, the Governor of Malta, lamenting over the death of his son, says

These arms of mine shall be thy sepulchre.

Compare 1. Henry VI. (Act IV, VII, 32), where Talbot, embracing the body of his son, exclaims:—

Now my old arms are young John Talbot's grave.

Compare, too, 3. Henry VI., Act II, IV, 114—5, where the Father says to the dead son whom he has killed in battle:

These arms of mine shall be thy winding-sheet;
My heart, sweet boy, *shall be thy sepulchre.*

Edward II.

was the best historical drama before Shakespeare.[2] This piece with its skilful play of motives and fine poetic portrayal of historic facts and personages must have been Shakespeare's constant model for his historical plays. Closer ties of relationship, especially, subsist between 'Richard II.' and 'Edward II.'

The following are noteworthy parallelisms:

[1] Comp. Shakesp. Jahrb. VI, 129 ff.

[2] Marlowe's best play is his Jew of Malta.

1) Edward II. Act I, IV, 407:

He wears a lord's revenue on his back

2. Henry VI. Act I, III, 83:

She bears a duke's revenues on her back.

2) In Edward II. Act II, II, a scene which contains the germ of a similar one in 1. Henry IV. (I, III), Mortimer declares (v. 125):—

Cousin, and if he will not ransom him,
I'll thunder such a peal into his ears,
As never subject did unto his king.

Compare Hotspur's language, *ut. sup.*, l. 219:

He said, he would not ransom Mortimer;
Forbad my tongue to speak of Mortimer;
But I will find him, when he lies asleep,
And in his ear I'll holla "Mortimer!"

3) Edward II., Act II, II, 162:

The wild Oneyl, with swarms of Irish kerns,
Lives uncontrolled within the English pale.

Compare the following verses from the 'First part of the Contention', which do not stand in 2. Henry VI:

Madame I bring you newes from Ireland,
The wild Onele my Lords, is up in Armes,
With troupes of Irish Kernes that uncontrold,
Doth plant themselves within the English pale.

4) The verse (Edward II., Act II, II, 166):—

The haughty Dane commands the narrow seas—

appears to be echoed in 3. Henry VI., Act I, I, 239:—

Stern Falconbridge commands the narrow seas.

5) Edward II., Act IV, III, 42—5:

Gallop apace, bright Phœbus, through the sky,
And dusky night, in rusty iron car,
Between you both shorten the time, I pray,
That I may see that most desirèd day,
When we may meet these traitors in the field.

Compare Romeo, III, II, 1—4, Juliet longing for the advent of night:—

Gallop apace, you fiery-footed steeds,
Towards Phœbus' lodging: such a waggoner
As Phaeton would whip you to the west,
And bring in cloudy night immediately.[1]

Doctor Faustus.[2]

I refer the reader to what I have said above on page 112.

No doubt from Marlowe's 'Faustus' Shakespeare derived many suggestions for his 'Tempest', in some respects the counterpart of the former. But Prospero and Faustus, synonymous though be their names, are very different characters. Both possess command over the natural and spiritual world; but how unlike their aims! Faustus mirrors the heaven-aspiring fiery spirit of the young author,—yea, even his violent and untimely end; while in Prospero is reflected the harmonious, serene, and gentle mind of the mature poet.

The following are parallel passages:—

1) Of 'Faustus', sc. IV, 65:

> *Wagner.* I will teach thee to turn thyself to anything; to *a dog, or a cat, or a mouse, or a rat* or anything.
> *Clown:* How! a Christian fellow to a dog or a cat, a mouse or a rat! etc.—

we have an apparent reminiscence in Romeo, III, I, 104:

> 'Zounds, *a dog, a rat, a mouse, a cat,* to scratch a man to death!

2) Dr. Faustus (sc. XIV, 83) receives Helen with the following words:

Was this the face that *launched a thousand ships*
And burnt the topless towers of Ilium?

Compare Troilus and Cressida, II, II, 82:

. . . . why, she [i. e. Helen] is a pearl,
Whose price hath *launch'd* above *a thousand ships.*

To the legend of Dr. Faustus, best known to the English public and Shakespeare through Marlowe's play, we find allusions in The Merry Wives,—Act I, I, 132:

How now, Mephostophilus!

---

[1] Compare also Tempest, IV, I, 29—31; and Henry V, Act IV, Prol. 20—23.— Mr. Fleay, in his edition of Edward II. (pp. 15—17), points out a number of similarities between 'Edward II. and Henry VI.'

[2] The play has come down to us in a corrupt state.

and Act IV, v, 70:

> [they] set spurs and away, like three German devils, three Doctor Faustuses.[1]

The Massacre at Paris.

The Duke of Guise's monologue (sc. II) deserves special notice as containing the probable germ of York's soliloquy in 2. Henry VI., III, I, 331 *seq.*

Dido,

written by Marlowe and Nash. The 'speech' recited in 'Hamlet' (II, II, 472 ff.) runs parallel to 'Dido', II, I, 181 ff.,—l. 255,

> And with the wind thereof the king fell down,—

corresponding to 'Hamlet', II, II, 496:

> But with the whiff and wind of his fell sword
> The unnerved father falls . .[2]

THOMAS KYD.

On Kyd's services to the English drama, I refer the reader to Boas's Introduction to the edition of the works of that early dramatist.

The Pre-Hamlet.

'That an old play on the subject of Hamlet existed there can be 'no doubt; it is referred to in 1589 .... by Nash, in his Epistle 'prefixed to Greene's Menaphon [cf. above, p. 36], and again in 1596, 'by Lodge (Wit's Miserie and the Worlds Madnesse, p. 56), where he 'alludes to "the Visard of *th*e ghost which cried so misera*b*ly at *th*e 'Theator like an oister wife, *Hamlet, revenge*".' 'Hamlet' is also mentioned as a play by Henslowe in 1594.[3] There is strong presumptive evidence in favour of the hypothesis, first advanced by Malone, that Kyd is the author of this pre-Shakespearean play. I must waive a discussion of this interesting and tempting problem

[1] Compare Bullen's ed. I, p. xxx note.

[2] Schröer, in his work on Titus Andronicus (p. 63) points out a coincidence between 'Dido' (Act III, sc. IV and Act IV, sc. I) and Titus Andronicus, II, III, 21 ff. —On 'Lust's Dominion', which was once unhesitatingly ascribed to Marlowe, see Schröer, *ibid.* Appendix.

[3] It must have been the Lord Chamberlain's men, who then played it at Newington Butts. They usually acted at The Theater in 1594—7.

here.[1] The following table gives the genealogy of the different versions of the story of Hamlet[2]:

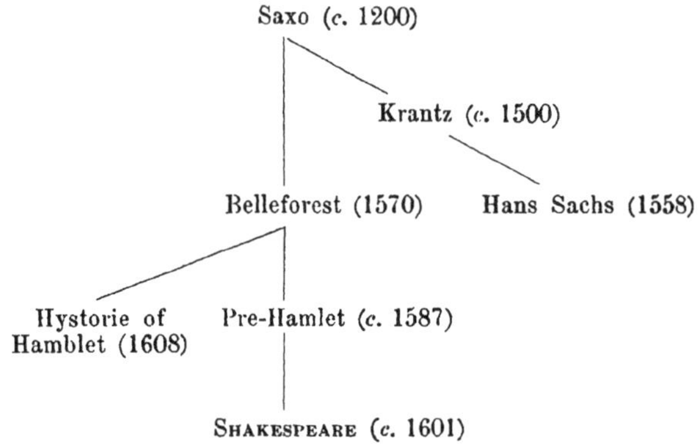

The next play we have to consider is

THE SPANISH TRAGEDY,

or the Second Part of Jeronimo,[3] one of the most popular of the early dramas on the pre-Shakespearean stage. Its theme of blood and revenge no doubt helped to suggest in large measure the horrible subject of 'Titus Andronicus' (see Boas, *ut sup.*, p. LXXIX).

The fifth scene of the second act, in which Jeronimo, alarmed by the cries of 'murder', enters hurriedly in his shirt and exclaims:—

> What out-cries pluck me from my naked bed,
> And chill my throbbing hart with trembling feare? etc.

---

[1] For further information, see Boas, *ut. sup.*, p. XLV, and Sarrazin, Thomas Kyd und sein Kreis. 1892. Mr. M. W. MacCallum, of the University of Sydney in Australia, cautions us not to be too hasty in our inferences (see English Miscellany presented to Furnivall. 1901. p. 282). Gericke and Moltke's 'Shakespeare's Hamlet-Quellen' (1881) is a handy little volume.

[2] The relation of the Bestrafte Brudermord (printed in 1781 from a MS. dated 1710) to Shakespeare is still under discussion. Mr. M. B. Evans has recently published a Doctor-Dissertation (1902) on this question, which is to be reprinted together with additional matter in Litzmann's Theatergeschichtliche Forschungen, vol. 19, which, I hear, will appear soon. Mr. Evans argues that the Bestrafte Brudermord is derived from Kyd's Pre-Hamlet,—for the reconstruction of which he is going to give us the materials. I cannot help entertaining some misgivings about the inferences he builds on rather slender foundations. Evans has Prof. Thorndike on his side (cf. Publ. of M. L. Ass., 1902, pp. 125 f.).

[3] A critical edition of The Spanish Tragedy was edited by Prof. Schick in 1901. I quote from Boas's complete edition of Kyd's works. Oxford, 1901.

enjoyed extraordinary celebrity and became the butt for unsparing persiflage and parody. It is alluded to by Shakespeare, in King Lear, III, iv, 48:—

Hum! go to thy cold bed, and warm thee—

and, more clearly, in The Taming of the Shrew, Ind., I, 9—10:—

Go by, *Jeronimy*: go to thy cold bed, and warm thee,—

the first portion of this latter quotation being derived from another passage in The Spanish Tragedy (III, xii, 31):

Hieronimo, beware; goe by; goe by.[1]

Perhaps there is also a sly allusion to the verse quoted above:

What out-cries pluck me from my naked bed?

in Mids. N. Dr., III, i, 132, where Titania, on awaking, exclaims:

What angel wakes me from my flowery bed?

The expression 'naked bed' is used by Shakespeare in 'Venus and Adonis', l. 397. Sly's *paucas pallabris* in the Shrew, Ind. I, 5, is a perversion of *pocas Palabras*, in The Spanish Tragedy (III, xiv, 118),—a phrase which had become a stock jest. Compare, too, Much Ado, III, v, 18:—

Dogberry: .... *palabras*, neighbour Verges.

From The Spanish Tragedy, II, i, 3, Shakespeare quotes in Much Ado, I, i, 263:—

'In time the savage bull doth bear the yoke'.[2]

The following are also probable reminiscences of The Spanish Tragedy. Titus Andron., IV, ii, 31:

But let her rest in her unrest awhile,

and Richard III., Act IV, iv, 29:—

Rest thy unrest on England's lawful earth.

Compare Span. Trag., I, iii, 5:

Then rest we heere a while in our unrest,

---

[1] 'Perhaps no single passage in Elizabethan drama became so notorious as 'this.' Boas, *ut. sup.* p. 406.

[2] Compare Furness's note. Instead of 'doth bear' Kyd has 'sustains'. Further allusions to Kyd's verse are in Much Ado, V, iv, 43: 'he thinks upon the 'savage bull'; and in Act V, i, 183: 'But when shall we set the savage bull's 'horns on the sensible Benedick's head?'

or Span. Trag., III, XIII, 29—30:

> Thus therefore will I rest me in unrest,
> Dissembling quiet in unquietnes.

The lamentations of Paris and Capulet at the supposed death of Juliet, in Romeo, IV, v, 58 and 62:

> O love! O life! not life, but love in death!
>
> O child! O child! my soul, and not my child!

remind us of Hieronimo's wail, after the death of his son, Span. Trag., III, II, 1:

> Oh eies, no eies, but fountains fraught with teares;
> Oh life! no life, but lively fourme of death;
> O world, no world, but masse of publique wrongs.

Perhaps King John, II, I, 137:

> You are the hare of whom the proverb goes,
> Whose valour plucks dead lions by the beard—

contains an echo from Span. Trag., I, II, 172:

> So Hares may pull dead Lyons by the beard.

'Margaret's lamentations (3. Henry VI., Act V, v) over the body of 'her own son Edward, the "sweet . . . plant . . . . . untimely cropp'd", 'echo the Marshal's (=Jeronimo) wail for his "sweet lovely Rose ill 'pluckt before" its "time"' (Boas).

Soliman and Perseda.

Basilisco, *the miles gloriosus* of 'Soliman and Perseda' (probably Kyd's play), who styles himself 'knight' but is called 'knave' in derision by the clown, is referred to by the Bastard, in 'King John', I, I, 244, where he, in reply to his mother's reproof

> What means this scorn, thou most untoward knave?—

says:

> Knight, knight, good mother, Basilisco-like.

The particular passage Shakespeare had in view is this (Act I, III, 165ff.):

> Piston, having got on Basilisco's back, makes him swear.
> *Bas.* O, I sweare, I sweare.
> *Pist.* By the contents of this blade—
> *Bas.* By the contents of this blade—
> *Pist.* I, the aforesaid *Basilisco*—

Bas. I, the aforesaid Basilisco—*Knight, good fellow, Knight, Knight*—
Pist. Knave, good fellow, Knave, Knave—etc.

When Shakespeare placed the following words on the lips of the Prince of Morocco (Merch. of Ven., II, I, 24):

. . . . By this scimitar
That slew the *Sophy* and a *Persian* prince
That won *three fields* of Sultan *Solyman*,—

he probably had in mind a passage from Soliman and Perseda, I, III, 51 ff., where the Turkish general brags:

Against the *Sophy* in *three* pitched *fields*,
Under the conduct of great *Soliman*,
Have I been chiefe commaunder of an host,
And put the flint heart *Perseans* to the sword.

In 'Soliman and Perseda' a fatal carcanet or necklace plays an important *rôle.* This chain—as Malone suggests, very plausibly—may have been in the poet's mind, when composing Othello. Compare, for example, Act III, IV, 55 ff.:

Oth . . . . That handkerchief
Did an Egyptian to my mother give . . . .
. . . . She, dying, gave it me;
And bid me, when my fate would have me wive,
To give it her. I did so: and take heed on't;
Make it a darling like your precious eye;
To lose't or give't away were such perdition,
As nothing else would match.

with the following passage from 'Soliman and Perseda', Act I, II, 32:

Perseda: . . . . accept this carkanet:
My Grandame on her death bed gave it me,
And there, ev'n there, I vow'd unto myselfe
To keepe the same, untill my wandring eye
Should finde a harbour for my hart to dwell.

Cinthio, from whom the plot of 'Othello' is taken, merely states that the handkerchief was presented to Disdemona by the Moor. (Furness, Othello, p. 220.)

JOHN LYLY.

Shakespeare also sat at the feet of 'eloquent and wittie'[1] Lyly, the writer of the best early comedies.

[1] Meres, Palladis Tamia, 1598, New Shaksp. Soc., Series IV, I, p. 161.

Lyly's brisk and witty dialogue, his fondness for conceits and puns, his use of prose as the vehicle for comedy, his graceful, elegant, and polished style, his skilful dramatic technique, his farcical scenes, his wit-combats among ladies and courtiers,—all this was imitated and seized upon by his alert disciple. Lyly's women, refined, witty, laughing, loving, or reserved, are the prototypes of many of Shakespeare's female characters. The idea of disguising girls as boys and the consequent mistaken identities were imitated by the greater poet. In Lyly's servants we discover the germ of some Shakespearean fools and clowns. Launce and Speed, for example, in The Two Gentlemen have much affinity with Licio and Petulus in 'Midas'. Lyly's mischievous little Cupido in 'Gallathea' and in 'Sapho and Phao' is Puck's forerunner. Dogberry and Verges of Much Ado have been compared with the watchmen of Endimion. Lastly, Lyly's fairies re-appear, in more beautiful garb, on Shakespeare's stage.

Shakespeare's first comedy, Love's Labour's Lost, is in direct imitation of Lyly's comedies, with which it shares the love-intrigue, the courtly atmosphere, covert allusions to court incidents, the skirmishes of wit, and the light vein. Armado and Moth have an unmistakeable resemblance to Sir Tophas and his page, in Endimion; while the Princess and her companions are as hard to please as Cynthia, but as sprightly and full of banter as many of Lyly's women and girls. 'The scene in Gallathea (III, i) where Diana's nymphs, 'entering one by one, confess their broken vow and agree to pursue 'their passion, has often been quoted as the original of that between 'the four anchorites, which is dramatically the best in Love's Labour's 'Lost' (Bond).[1]

The following detailed resemblances are worth noticing:—

Endimion.

The fairies, who, in The Merry Wives, V, v, pinch Falstaff while they sing:—

[1] Compare Brandl, "Shakspere" p. 44.

Love's Labour's Lost has often been designated a *Tendenz*-drama, a *précieuses ridicules*' of those days, written with the purpose of ridiculing 'four chief 'affectations in speech'. I cannot accept this view. Shakespeare does, I admit without hesitation, mock extravagances of style, but only incidentally. I am of the opinion that *Johannes factotum*, who had written the earliest tragedies and histories in distinct imitation of Kyd and Marlowe, did no more than follow Lyly as his model, in Love's Labour's Lost. The taffeta phrases which Biron renounces in Act V, ii, 406, refer to the affected language he used as a wooer (Comp. V, ii, 34 ff. and 787—794).

Pinch him, fairies mutually;
Pinch him for his villany;
Pinch him, and burn him, and turn him about ....

are following the example of the fairies of 'Endimion' (IV, III), who punish Corsites in the same manner, while singing:

.... Pinch him, pinch him, blacke and blue, ...
Pinch him blue.
And pinch him blacke, etc.

Compare also similar verses, The First Song, in Endimion, III, III. Note, that both of the 'pinch'-songs are in the four-beat measure.

Campaspe.

Of the following passage in Campaspe, II, II, 35ff.:

Is the warlike sound of drumme and trumpe turned to the soft noyse of lire and lute? the neighing of barbed steeds, whose loudnes filled the ayre with terrour, and whose breathes dimmed the sunne with smoak, converted to dilicate tunes and amorous glaunces, etc?

we have an apparent reminiscence in Richard III., Act I, I, 7ff.:—

Gloucester: Our stern alarums [are] changed to merry meetings,
Our dreadful marches to delightful measures.
Grim-visaged war hath smooth'd his wrinkled front;
And now, instead of mounting barbed steeds
To fright the souls of fearful adversaries,
He capers nimbly in a lady's chamber
To the lascivious pleasing of a lute.
But I, that am not shaped for sportive tricks,
Nor made to court an amorous looking-glass ....

The following lines in Cymbeline, II, III, 21:

Hark, hark! the lark at heaven's gate sings
And Phoebus 'gins arise.

seem to echo Campaspe, V, I, 37ff.:

.... *the Larke* so shrill and cleare;
How *at heavens gats* she claps her wings,
The Morne not waking till shee sings.
Heark, heark, ...

Compare, too, Sonnet XXIX:

Like to the *lark* at break of day arising
From sullen earth, sings hymns *at heaven's gate.*

Again, Campaspe, III, II, 37-8:

> .... for *thy dull head* will bee but a *grindstone for my quick wit,* which if thou whet with overthwarts, *periisti*—

was perhaps in Shakespeare's mind when he wrote As You Like It, I, II, 57-9:

> [Nature] hath sent this natural for our whetstone; for always *the dulness of the fool is the whetstone of the wits.*

In Act III, sc. v, 37, of 'Campaspe' Apelles, enamoured of the heroine of the play, on whom Alexander the Great has cast his eye, exclaims:

> starres are to be looked at, not reched at.

Similarly, the duke of Milan, in 'The Two Gentlemen', III, I, 156, who is desirous of giving his daughter to the rich Thurio, addressing Valentine, the rival lover, says:

> Wilt thou reach stars, because they shine on thee?[1]

Mother Bombie.

In Act IV, II, 28, the half-witted Silena mistaking Accius for a stool, says:

> I crie you mercy, I tooke you for a ioynd stoole.

The same mistake and the same apology is made by the Fool in King Lear, III, VI, 54. Compare, too, The Shrew, II, I, 199.

In Act I, III, of 'Mother Bombie' Candius has a stolen *rendez-vous* with Livia, whom he teaches the art to love:

> *Livia*: What booke is that?
> *Cand.* A fine pleasant poet, who entreateth of the *art of Love,* and of the remedie.

Candius then quotes apposite verses from Ovid's Ars Amatoria. Similarly, Lucentio, in The Shrew reads "the Art to Love" with Bianca and quotes and construes Ovid.

The Woman in the Moon.

Puck's apology to the public, in the Epilogue to Mids. N. Dream:

[1] But the same thought occurs in Greene's Pandosto, 1588 (see Furness, XI, p. 342): "starres are to be looked at with the eye, not reacht at with the hande." Compare Goethe's lines in his exquisite poem, beginning 'Wie kommt's, dass du so traurig bist':—

> Die Sterne, die begehrt man nicht,
> Man freut sich ihrer Pracht.

Mrs. v. Koenen drew my attention to this.

If we shadows have offended,
Think but this, and all is mended,
That you have but slumber'd here
While these visions did appear.

is anticipated in Lyly's Prologue to 'The Woman in the Moon':

If many faults escape in her discourse,
Remember all is but a Poets dreame.

Compare also the Prologue before Sapho and Phao.

Many more correspondences between Lyly and Shakespeare (some of which are doubtful) are pointed out by Mr. Bond, in his edition.

The influence of the remaining dramatists is less marked and generally eludes precise measurement. The following, of whom we find more or less clear traces in Shakespeare, require special mention.

GEORGE PEELE.

The following words, placed on the lips of ranting Pistol, in 2. Henry IV, Act II, IV, 193:

Then feed, and be fat, my fair Calipolis,—

allude burlesquingly to a scene in Peele's BATTLE OF ALCAZAR (Act II, sc. III), where Muly Mahamet, presenting a piece of 'lion's flesh' on the point of his sword to his wife, says:

Feed, then, and faint not, fair Calipolis
(vv. 81 and 94)—

and again, v, 101:

Feed and be fat, that we may meet the foe.

The following verses at the end of the next following scene of Peele's play:

Saint George for England! and Ireland now adieu,
For here Tom Stukeley shapes his course anew—

are closely paralleled in King Lear, I, I, 188:

Thus Kent, O princes, bids you all adieu;
He'll shape his old course in a country new.[1]

Pistol's

Have we not Hiren here?

---

[1] On the Moor's exclamation in Alcazar (Act V, sc. 1): "A horse! a horse, "villain, a horse", see Dr. Churchill's excellent observation in 'Richard III. up to Shakespeare', (forming the 10th volume of 'Palaestra'. Berlin, 1900).

in 2. Henry IV., Act II, IV, 173 and 189, is supposed to be a quotation from Peele's lost play: *The Turkish Mahomet and Hiren the Fair Greek.*[1]

ROBERT GREENE.

Several coincidences, though not very striking, between Shakespeare's and Greene's plays are pointed out by Sarrazin.[2] There is much family likeness between Greene's 'Selimus' (*ca.* 1587) and Titus Andronicus, two plays of appalling horrors and brutalities. Dowden, in his Primer (p. 7), lays stress on the romantic notes sounded in Greene's dramas. On the whole, Shakespeare's indebtedness to his envious rival does not seem very marked.[3]

GEORGE WHETSTONE.

The general outlines of Shakespeare's 'Measure for Measure' together with hints for some of its comic figures belonging to a very low grade of society were supplied by Whetstone's play of Promos and Cassandra printed in 1578. The plot of the serious portion of Whetstone's play agrees in substance with the prose narrative by him, entitled 'The Rare Historie of Promos and Cassandra. Reported by Madam *Isabella*,' inserted in his 'Heptameron of Civill Discourses', 1582. The name Isabella suggests that the prose narrative was also known to our poet. Whether Shakespeare did, or did not, know Cinthio's novel, (Hecatommithi, Nov. 5, Dec. 8.) the original of the story, he does not appear to have used it.[4] Nor does he seem to be indebted to Cinthio's Epitia, a play on the same subject.[5]

[1] Compare Creizenach, *ut sup.*, p. XXXII; and Cohn, Sh. in Germany, p. LXVII; and Murray's New Engl. Dict, s. v. Hiren.— Mr. Fleay's supposition (Chronicle of the Engl. Drama, II, 157), that the following lines from Peele's *Edward I:*

"Shake [thou] thy spears in honour of his name,
"Under whose royalty thou wear'st the same"—

imply that the part of Edward was acted by Shakespeare, seems to me a very hazardous guess. Nor is there the slightest connexion between Shakespeare's name and a passage in Lyly's 'Campaspe' (see Klein, XIII, 525).

[2] Shakespeare's Lehrjahre, see his index.

[3] Cf. Englische Studien, vol. 22, pp. 389 ff. On Selimus see H. Gilbert's Dissertation, Kiel, 1899; and Shakesp. Jahrb., XXXVIII, 297.—Mr. Ch. Collins, I may remark, is occupied upon an edition of Greene's plays, to be issued by the Clarendon Press.

[4] See Shakesp. Jahrb. XIII, 169—170.

[5] See Klein Gesch. des Drama's, V, 354; cf. above, p. 71.

GEORGE GASCOIGNE.

The disguises of Lucentio, Hortensio, and Tranio, the character of old Gremio, the engaging of the pedant, his personation of Vincentio, etc.—for which we find only a few faint hints in 'The Taming of *a* Shrew'—are derived from Gascoigne's SUPPOSES (1566), translated from Ariosto's 'I Suppositi'.[1] The names Petrucio and Lytio reappear in Shakespeare's play as Petruchio and Licio; and in The Shrew, V, I, 120—

> While counterfeit *supposes* blear'd thine eyne—

we have a probable reminiscence of the title of Gascoigne's drama.

THOMAS PRESTON.

To Preston's CAMBYSES (*circa* 1569) Falstaff alludes in 1. Henry IV., Act II, IV, 425:

> I must speak in passion, and I will do it *in King Cambyses' vein.*

Falstaff then continues:

> Weep not, sweet queen; for trickling tears are vain.
> .... For God's sake, lords, convey my tristful queen;
> For tears do stop the flood-gates of her eyes—

which may be in allusion to a scene towards the end of 'Cambyses', where we read (Dodsley, IV, 236):—

> [At this tale let the Queen weep.
> *Queen.* These words to hear makes stilling tears
> Issue from crystal eyes.
> *King.* What dost thou mean, my spouse, to weep
> For loss of any prize?

Notwithstanding the above persiflage Shakespeare seems not to have disdained to borrow a *motif* from Preston's play. Neither in Whetstone's drama of Promos and Cassandra nor in his prose account of the story does the king leave the town and appoint a deputy, who abuses this office of trust. This may have been suggested to the poet by Preston's drama.

BEN JONSON.

Shakespeare took part in the original performances of Jonson's 'EVERY MAN IN HIS HUMOUR' (1598), and his 'SEJANUS' (1603). According to Mr. S. Lee "a prologue arm'd", in 'Troilus and Cressida',

[1] Compare, too, Farmer's remarks (Malone's Var. Ed., vol. I, p 340-1).

Prol. 23, alludes to the 'armed prologue' of Jonson's POETASTER (1601).[1] The author of the Epilogue to 'Henry VIII.', who wrote

> others [come] to hear the city
> Abus'd extremely, and to cry,—"That's witty!"—

was possibly referring to Jonson's satiric efforts.

Jonson's 'humours' and his realism seem to be reflected in some measure in Shakespeare's Merry Wives (*c.* 1598). Nym's harping upon the word 'humour' in this play as well as in Henry V. is very noteworthy.

It is probably not a matter of mere accident, that the names Prospero and Stephano should occur both in Jonson's first-named play (Quarto ed., 1601[2]) and in Shakespeare's Tempest. Farmer[3] even maintained that Ben Jonson taught Shakespeare the correct pronunciation of the latter name, pronounced Stéphano in The Tempest, but Stepháno in the Merchant of Venice, as the metre shows. Elze thinks the metre does not show this.[4]

SAMUEL ROWLEY

had written a drama on Henry VIII., entitled '*When you see me, you know me, or the famous Chronicle Historie of King Henrie VIII*'., and printed in 1605. It was republished by Karl Elze in 1874. 'Fletcher 'and Shakespeare', says Mr. Lee in his article on Rowley in the Dict. of Nat. Biogr., 'possibly owed something to Rowley's effort when preparing 'their own play of Henry VIII.' Shakespeare's drama was apparently a rival production of Rowley's work, which, as Boswell[5] pointed out, is the probable object of attack of the Prologue of the former play.

FLETCHER AND BEAUMONT.

Tennyson, the great Victorian poet, was the first to start the theory, which is gaining more and more ground, that Fletcher was a collaborateur of 'Henry VIII'.

Professor Thorndike, in his work 'The Influence of Beaumont and Fletcher on Shakspere, 1901', points out numerous resemblances between Beaumont and Fletcher's PHILASTER (1609-10?), acted by

[1] Cf. S. Lee, Life of Shakespeare, 1899, pp. 44 and 229.
[2] Republished by C. Grabau, in Sh. Jahrb. XXXVIII.
[3] See Furness, IX, 4.
[4] Notes on Elizabethan Dramatists, 1889, p. 160.
[5] Cf. Malone's Var. Ed., vol. XIX, p. 501; and Elze's Edition.

Shakespeare's company, and Shakespeare's Cymbeline. 'It may be, 'indeed (says Prof. Dowden[1]) as Professor Thorndike contends, that '*Cymbeline* was influenced by *Philaster*; but, on the other hand, there 'is no decisive evidence to show that *Cymbeline* was not the earlier 'of the two plays.' May one venture to hazard the conjecture that Beaumont or Fletcher, who wrote for Shakespeare's company, may have had a hand in a Pre-Cymbeline?

Further, Prof. Thorndike argues that The Winter's Tale and The Tempest possess many of the characteristics of the Beaumont-Fletcher romances. But the type of the Shakespearean dramatic romance had been at least outlined or sketched in *Pericles.*[2]

SIR WILLIAM ALEXANDER,

Earl of Stirling, a respected politician and poet at the court of King James.

There is a noteworthy parallelism between Stirling's DARIUS (1603), Act IV, II:

> Let greatnesse of her glascie scepters vaunt;
> Not sceptours, no, but reedes, soone brus'd, sone broken:
> And let this worldlie pompe our wits enchant.
> All *fades*, and *scarcelie leaves behinde a token.*
> Those *golden Pallaces*, those *gorgeous* halles,
> With fourniture superfluouslie faire:
> Those statelie Courts, those sky-encountring walles
> *Evanishe all like vapours in the aire.*

and The Tempest, IV, I, 148,

> These our actors,
> As I foretold you, were all spirits and
> Are *melted into air,* into thin air:
> And, like the baseless fabric of this vision,
> The cloud-capp'd towers, the *gorgeous palaces,*
> The solemn temples, the great globe itself,
> Yea, all which it inherit, shall dissolve
> And, like this insubstantial pageant *faded,*
> *Leave not a rack behind.*

Similar thoughts with less verbal resemblance will be found in Spenser's Ruins of Time, 1591, (vv. 91—'9) and in Ruins of Rome. The thought is, of course, biblical.

[1] Cymbeline ed. by Dowden, 1903, p. XIII.
[2] Dowden, *ut sup.*, p. XV.

Alexander's 'Julius Caesar' is chronologically later than Shakespeare's play on the same subject.

ANONYMOUS PLAYS RETOUCHED BY, OR KNOWN TO, SHAKESPEARE.

We now proceed to a brief discussion of anonymous plays, both extant and hypothetical, which were rehandled by the great dramatist, after which we shall take note of other plays and dramatic productions reflected in Shakespeare's works. As regards the old plays rehandled by Shakespeare, it will be observed that several of them belonged to the Queen's Men, who, says Mr. Fleay, evidently broke in 1591. They had played at the 'Theater', which was occupied for a number of years by Shakespeare's company,—called successively, Lord Leicester's, Lord Strange's, Lord Derby's, Lord Chamberlain's[1] (1594—1603), and the King's Men (1603—1642). Many of the plays of the Queen's Men are traceable to Shakespeare's troupe and the Earl of Sussex's. The latter disappear about 1594.[2]

THE TRUE CHRONICLE HISTORIE OF KING LEIR

is probably identical both with the play of 'king leare',[3] mentioned by Henslowe on the 6th and 8th of April 1594, and with the 'Chronicle historye of Leire', entered on the Stationers' Registers on 18 May 1594. But the only edition preserved to us is dated 1605.[4] This play, there can be little doubt, was the chief source of Shakespeare's immortal King Lear. The question was carefully discussed by Mr. Furness, in his New Var. Ed. (V, pp. 383 ff.), where an account with the contents of the old play will be found.

Though Shakespeare followed this old play as his main source he must have made further studies in the Lear-story. First, the tragic end of Shakespeare's play was not derived from the old drama, which ends happily, but from some other version containing an ac-

[1] Temporarily also Lord Hunsdon's Men.

[2] See Mr. Fleay's History of the London Stage.

[3] Entered as one of the plays 'by the queens men and my lord of Sussex' 'together'. (Malone, III, 301.)

[4] No statement is made as to the company to which it belonged.

count of Cordelia's death. Further, Shakespeare's play has features in common with the original version, which are not to be found either in Spenser, or in the old play, or in Holinshed. Geoffrey of Monmouth mentions the reason, or rather the occasion, of the rupture between parent and the two daughters: the gradual diminution of his retinue of sixty knights (Shakespeare has a hundred). This is not referred to, or only very faintly hinted at, in the other versions just mentioned.[1] But it is to be found in the 'Mirror for Magistrates'.[2] Another point of agreement between Shakespeare and the Geoffrey (-Mirror) version is this. In his play *Goneril* is the wife of the *Duke* of *Albany*, where Lear sojourns first. In the old drama she is the wife of the *King* of *Cornwall*. Holinshed's words are not clear on this point at all. In Geoffrey Gonorilla is married to the '*dux Albaniae*', in the Mirror to the '*Prince* of *Albany*'.

To Camden's Remains Shakespeare is said to be indebted for Cordelia's reference to conjugal love in her fatal answer (Lear, I, I, 98ff.). Apart from Camden, only Polydore Vergil and the Mirror have a like reference. However, the coincidence may be accidental. It is not certain whether Camden's Remains concerning Britain (1605) appeared anterior to King Lear.[3]

From Spenser's Fairy Queen (II, x, 27ff.) Shakespeare took the name Cordelia (see *ante*, p. 90).

THE TROUBLESOME RAIGNE OF KING IOHN,

acted by the Queen's Men and printed in 1591 is the original of Shakespeare's King John, which follows it very closely, almost scene for scene.

The author of the older play, I may note parenthetically, probably derived most of his materials from Holinshed, without adhering

[1] Geoffrey's version is followed by Robert of Gloucester, Robert of Brunne, and Layamon. In the Gesta Romanorum, too, Lear has 40 knights, which number is gradually reduced.

[2] But the Mirror gives no clear reason for the dismissal of the 60 knights, while Geoffrey distinctly refers to the quarrels of the knights with the servants and members of the households. Similarly, Shakespeare's Goneril complains of the "rank and not-to-be-endured riots" and quarrels of Lear's retinue. This coincidence may or may not be accidental.

[3] Concerning the Lear-story I refer to Eidam, Die Sage von K. Lear. Programm, Würzburg, 1880. It is a matter of regret that he made no attempt at forming a genealogical table of the different versions.

to historical truth. (See Boswell-Stone, Shakspere's Holinshed.) He must have also referred to Grafton's (or Caxton's) Chronicle for details concerning the death of the king. 'The inwards of a Toad', for example, to poison the king, as well as the absolution of the monk before committing the crime, are not taken from Holinshed.

Though Shakespeare may have referred to Holinshed, he makes no independent use of this work.[1] One item in his play betrays that he had opened his Grafton.[2]

THE FAMOUS VICTORIES OF HENRY THE FIFTH,

printed in 1598,[3] but acted much earlier, probably even before 1588, by the Queen's Men, and entered on the Stationers' Registers in 1594, May 14. The 'Harry V', mentioned in Henslowe's Diary as being acted (probably) by the Admiral's Men in 1595—96, may be identical with the Famous Victories.[4]

The popularity which this old play probably enjoyed was doubtless the occasion of the composition of Henry IV. and Henry V. Vastly inferior though it is to these, it has the honour of having presented in dramatic form the bare outlines of the story with numerous hints of Shakespearean scenes, and the figure of Henry, first as the madcap Prince amongst his wild companions (of whom Sir John Oldcastle is one), and then as the good king and triumphant conqueror of France.

THE TRUE TRAGEDIE OF RICHARD THE THIRD.

It appears that Shakespeare was under obligations to this play, acted by the Queen's Players and printed in 1594 (but of older date),

[1] The death-scene in King John (V, vii, 49—65) reminds one of an account related by Holinshed (omitted by Boswell-Stone), according to which the king's death was accelerated by an evil tiding. In the older play he receives only consolatory news immediately before his death.

[2] "A monk, I tell you; a resolved villain,
"*Whose bowels suddenly burst out.*"
(John, V, vi, 29)

Compare Grafton (ed. 1809) p. 246: 'The Monke anone after went to the Farmory, and there dyed, his guttes gushing out of his belly'. In Caxton's Chronicle we read 'his wombe was broken in sonder'. (Comp. my Dissertation, Berlin, 1900.)

[3] Facsimiled by Praetorius, with an Introduction by P. A. Daniel, in 1887.

[4] Collier's "harey the vth" of 1592 is either a blunder or a forgery. It is high time that Henslowe's Diary were re-edited.

or to some play nearly related to it, for the conception and some traits and phrases of his 'Richard III.'[1]

THE TAMING OF *A* SHREW

had been acted by Lord Pembroke's Servants as we learn from the title page of the play, as printed in 1594. But the play must have passed into the hands of the Lord Chamberlain's Men in or before 1594, as we may infer from Henslowe's mention of it among the plays acted by the Admiral's and the Chamberlain's Men at Newington Butts (South of the Thames) from June 3 to 13, 1594.

Some commentators maintain, that Shakespeare refurbished the old play (which contains the general outlines of The Taming of *the* Shrew) after it had been recast by another hand.[2]

A PRE-TIMON.

Professor Herford, pointing out the resemblances between Shakespeare's play of Timon and Lucian's Dialogue *Timon*, concludes: 'Lucian's dialogue evidently comes nearer to the drama than either 'Plutarch or Painter [comp. my note above on p. 41]. The entire 'scheme of the plot is already there, and the germ of Timon's character.'[3] In what form Lucian's dramatic dialogue reached Shakespeare is not known. For my part, I am inclined to believe that he had an older play before him which he revised, bringing out into clearer relief the character of Timon, one of Shakespeare's most powerful creations.[4] The relation between Shakespeare's drama and a manuscript academic play printed in 'Shakespeare's Library' is well discussed by Prof. Brandl.[5] Whether it was, however, known to Shakespeare remains uncertain.

[1] See Dr. Churchill's dissertation on Richard III. up to Shakespeare (Palaestra X, Berlin 1900),—a work seen through the press with patience, care, and self-denial by my friend Mr. W. Perrett. Compare also Prof. Brandl's succinct introduction to the play, in his edition of Schlegel-Tieck's translation of Shakespeare. Dr. Churchill's remarks relating to Henslowe's Play on Richard III. require to be corrected according to Fleay's Chronicle of the English Drama, vol. II, p. 284, item 13.

[2] Comp. Dr. Furnivall, New Shaksp. Soc. Trans., 1874, p. 102—114; and also Herford, Eversley Ed., vol. 2.

[3] From Herford's able Introduction, Eversley Edition, X, p. 155.—Cf. also Lloyd's Essay.

[4] An estimable lady, who saw Shakespeare's 'Timon' acted by the Meiningen Company, has repeatedly told me that the play made an ineffaceable impression on her mind.

[5] Schlegel-Tieck Translation, VI, 235-8.

A PRE-MERCHANT.

The hypothesis of a pre-Shakespearean play on the same subject as his Merchant of Venice is based on the following passage in Stephen Gosson's School of Abuse (1579):[1]

> And as some of the Players are farre from abuse, so some of their Playes are without rebuke: which are as easily remembered, as-quickly reckoned. The twoo prose Bookes plaied at the Belsavage, where you shall finde never a woorde without wit, never a line without pith, never a letter placed in vaine. *The 'Iew'* and 'Ptolome', showne at the Bull, *the one representing the greedinesse of worldly chusers, and bloody mindes of Usurers:* The other very lively discrybing howe seditious estates, with their owne devises, false friendes, with their own swoordes, and rebellious commons in their own snares are overthrowne: neither with Amorous gesture wounding the eye: nor with slovenly talke hurting the eares of the chast hearers.

There is a general consensus of opinion that 'The Jew' referred to by Gosson is the forerunner of Shakespeare's play. From 'the 'greediness of worldly choosers and the bloody minds of usurers' we infer that it combined the story of the caskets and that of the pound of flesh.[2]

What the 'venesyan comodey' of Henslowe's Diary (1594) was, and whether there is any link of connexion between it and Shakespeare's play, must for ever remain dubious.

May we suppose that Spenser alludes to the old play in a humorous letter to Harvey, dated 1579 (?), where he signs himself thus:

> He that is faste bownde unto the in more obligations then any marchante in Italy to any Jewe there.

And did Greene, too, make an allusion to the casket story of the old play, in 'Mamilia' (1583)?—

[1] Arber's Reprint 1868, p. 40.

[2] There is one dissentient voice. Mr. Fleay, who is entitled to a respectful hearing, seeks to identify Gosson's *Jew* with *The Three Ladies of London* (printed 1584). Moreover he suggests that 'chusers'=chousers. (Hist. of the Stage, p. 40.) What Dr. Murray says about 'chouse(r)' in his Dictionary seems to tell against Mr. Fleay's interpretation of the word. It is difficult to see why 'The Three Ladies of London' should have ever acquired the name of the *Jew*, as the Jew is not at all the prominent figure of the play: Moreover the Jew is not *Jewish.*

He which maketh choyce of bewty without vertue commits as much folly as Critius did, in choosing a golden boxe filled with rotten bones.

But it would be unsafe to build conclusions on these two passages.[1]

A PRE-GENTLEMEN.

The story of the shepherdess Felismena, related in Montemayor's 'Diana', had been dramatized as early as 1585 in a play entitled 'The History of Felix and Philismena', acted at Greenwich before the Queen. Now, Don Felix in the 'Diana' corresponds to Proteus in The Two Gentlemen of Verona, while Felismena is Shakespeare's Julia. There is therefore much probability in the supposition that Shakespeare was indebted to a dramatic version of Montemayor's tale for the main incidents of his 'Two Gentlemen'.[2]

FURTHER OLD PLAYS,

REAL, HYPOTHETICAL, AND CHIMERICAL,—SUPPOSED (RIGHTLY OR WRONGLY) TO HAVE FORMED THE BASES OF SHAKESPEAREAN PLAYS.

In this section dealing with purely hypothetical matter, which requires further working out, I mention briefly: I) hypothetical plays which I consider likely sources of some Shakespearean plays, though the external evidence is somewhat meagre; II) old plays, really existent, but probably not the bases of Shakespearean plays; III) chimerical or unlikely sources,—plays that have been suggested on very insufficient grounds to have formed the bases of the respective Shakespearean plays.

I) HYPOTHETICAL PLAYS, BEING LIKELY SOURCES.

Shakespeare's Othello, Cymbeline, Much Ado, and The Merry Wives are founded on Italian tales, of which English translations do not seem to have existed. This circumstance, the marked divergence of these plays from their Italian originals[3], Stephen Gosson's assertion that numerous stories of all sorts had been dramatized to furnish the play houses of London (cf. *ante*, p. 59)—an assertion which extant lists of the names of old plays strongly bear out—, Shakespeare's apparent want of knowledge of Italian literature beyond the stories

[1] Comp. Furness, VII, p. 322; Englische Studien, XVI, 372.

[2] Comp. *ante*, p. 72.

[3] 'Othello' keeps more closely to its source.

forming the plots of some of his dramas[1], his practice of retouching old plays in other known cases,—all this *seems to me* to point to older plays as the bases of his respective dramas just mentioned. It would be too much to suppose that a mortal human being, so busy in the practical affairs of life, who had to act in the play-house every day, and who was one of the leaders of his company, could have written all the plays, the best ever produced, single-handed within the short space of about twenty years.

A PRE-CYMBELINE.—The crudeness of some scenes and the unequal execution of the play suggest another hand.

A PRE-OTHELLO.—'To some readers it will seem likely that 'Shakespeare had some other or fuller version of the story than '[Cinthio's] as his material.' These are the words of Mr. H. C. Hart (in his edition of Othello, 1903, p. XXXI), who finds evidence in support of his view in the absence of any of Shakespeare's names in Cinthio except Desdemona, in Cassio's commercial pursuits (I, I), in the names 'Sagittary' and 'Marcus Luccicos' (I, III), and in the general trend of the historical events in Othello. I would also draw attention to the poet's knowledge of Venetian affairs and customs.

Further, I hypothetize a PRE-MUCH-ADO,[2] a PRE-TWELFTH-NIGHT (cf. *ante* pp. 65—70), and a PRE-MERRY-WIVES. Fleay's supposition, that the 'gelyous comodey' mentioned by Henslowe, Jan. 5, 1593, is connected with the Merry Wives, is very hazardous. As to a PRE-ERRORS, see *ante*, p. 32f.[3]

PERICLES is only in part by Shakespeare. Whether he retouched a play already finished by another hand, or whether he collaborated with another author, we cannot tell with certainty. I am inclined to accept the former supposition. In this sense, I believe in a Pre-Pericles, which, however, suffered little change under Shakespeare's hands. The hypothesis of a Pre-Pericles of an *early* date rests on spurious and forged data (cf. *ante*, p. 81, note 2).

[1] Titus Andronicus, too, is partly based on a tale of Bandello's (see p. 66).

[2] Mr. Furness makes the same supposition, Var. Ed., XII, xx; though his conjecture about 'Benedicte and Betteris', mentioned in 1613, does not help us at all.

[3] Is *All's Well* based on an older play?

II) REALLY EXISTENT OLD PLAYS, BUT PROBABLY NOT BASES OF SHAKESPEAREAN PLAYS.

We know that there existed older English plays on the subject of Julius Caesar, Antony and Cleopatra, Troilus and Cressida, and perhaps also on Romeo and Juliet. Whether Shakespeare was indebted to them is open to doubt. It is hardly necessary to say, that precedence in time does not involve dependence.

A PRE-ROMEO?—In the preface to his poem of Romeus and Juliet (1562), the basis of Shakespeare's 'Romeo and Juliet', Arthur Brooke says: 'Though I saw the same argument lately *set foorth on stage* with more 'commendation, then I can looke for: (being there much better set forth 'then I have or can dooe) yet the same matter penned as it is, may 'serve to lyke good effect'. It is difficult to see, why the expression, 'set foorth on stage', should be taken in a figurative sense, as Delius suggested.[1] But whether this play was an English play, and whether it, or some English play succeeding it or actually founded upon it, was known to Shakespeare, is a question which is for ever doomed to remain matter for conjectural speculation only. Mr. P. A. Daniel thinks he can discover traces of a Pre-Romeo in the first Quarto edition. But he admits that it is impossible to give satisfactory evidence.

JULIUS CAESAR had, of course, appeared on the pre-Shakespearean stage. A Latin play on Caesar's death, of which Richard Edes was the reputed author, had been acted at Christ Church in Oxford in 1582.[2] It is very likely that Henslowe's 'Caesar and Pompey', 1594-5, consisting of two parts, differed but little from 'The Tragedie of Caesar and Pompey or Caesar's Revenge', printed in 1607 (and,

[1] Comp. Brooke, v. 2817:

'The prince did straight ordaine, the corses that were founde
'Should be *set forth upon a stage* hye raysed from the grounde.'

Here the expression is certainly *not* used in a figurative sense.

[2] Dr. A. Ward, in his Hist. of Engl. Dram. Lit., II, p. 141, says that the famous *et tu Brute* of the murder-scene occurred in Edes's play. I suspect Dr. Ward accepted as a fact what Malone had only stated conjecturally. But the phrase must have been pre-Shakespearean. It is placed on the lips of Edward in The True Tragedie of Richard Duke of York (printed in 1595): "Et tu, Brute, wilt thou stab Caesar too?" The expression occurs also in Jonson's 'Every Man out of his Humour', printed in 1600, Act V, IV.

possibly, earlier),—a Senecan play, with some Marlowian touches. It ends with the death of Brutus. Henslowe's play is probably one with, or is an enlarged version, of 'Caesar and Pompey' mentioned by Stephen Gosson, 1582.[1] Probably it was known to Shakespeare, whose play is, however, based on Plutarch. To representations of Julius Caesar on the stage we have two noteworthy allusions in Hamlet, III, II, 104f., and Julius Caesar, III, I, 111—4.

Alexander's Julius Caesar, printed in 1607, is later than Shakespeare's play. According to Collier's edition of Henslowe, Drayton and others were occupied upon a play called 'Caesar's Fall' in 1602. This would also be later than Shakespeare's.

Older Plays on Antony and Cleopatra.—Neither to the Countess of Pembroke's Translation of Garnier's *Marc Antoine*, 'Antonius, a Tragedy', first printed in 1592, nor to Samuel Daniel's 'Cleopatra' (1594), a companion-piece to this tragedy, nor to Samuel Brandon's 'Virtuous Octavia' (printed 1598), does Shakespeare owe anything for his 'Antony and Cleopatra' based on North's Plutarch.

A play on the subject of Troilus and Cressida was written by Dekker and Chettle in (or very soon after) 1599, as we know from Henslowe's Diary. Shakespeare must have heard of it, for his play is apparently a rival production. But the loss of the former renders all speculation on this point fruitless. The Troilus-play entered on the Stationers' Registers on 7. Febr., 1602—3, as a play 'acted by 'my lord Chamberlens Men' must have been Shakespeare's.

2. and 3. Henry VI.—English and German critics are at variance in their views with regard to Shakespeare's share in the composition of 2. and 3. Henry VI. While the former, led by Malone, agree in holding that 'The First Part of the Contention' and 'The True Tragedy of Richard Duke of York' were not, or not wholly, by Shakespeare, who recast them, forming 2. and 3. Henry VI out of them;—the latter, led by Delius, suppose that those plays are imperfect and

[1] Compare Aldis Wright, Clarendon Press Edition of Caesar, p. IX. Mr. Lee, in his Life of Shakespeare, p. 211, says: 'A play of the same title [Julius Caesar] 'was known as early as 1589, and was acted in 1594 by Shakespeare's company'. Mr. Lee is referring to Henslowe's play, the title of which is however different. His date 1589 is probably an error for 1582. We have no evidence that Shakespeare's company acted the play in question (cf. Fleay, Hist. of the Stage, p. 140).

pirated copies of Shakespeare's plays. The question ought to be rediscussed with an unbiassed mind.[1]

III) CHIMERICAL AND UNLIKELY SOURCES.

Regarding the hypothesis of a PRE-TITUS-ANDRONICUS, the chief facts which require to be attended to are these:

1) A play 'titus and ondronicus' (probably=Titus Andronicus) is mentioned in Henslowe's Diary on 23 Jan., 1594, and marked *ne* (=new play). Now, this *ne* (whether it is authentic or not) only means that the play was then first acted by Henslowe's troupe. The *ne* has therefore little significance for us.[2]

2) Critics have supposed that there existed an earlier version of 'Titus' which they believe was identical with 'tittus and Vespacia' (=Vespasian), mentioned by Henslowe in 1592, for the reason that the German version of Titus Andronicus (printed 1620), in which the names are freely altered, Titus's eldest son is called Vespasianus. But it is much more likely that Henslowe's 'Titus and Vespasian' was a play based on Josephus (cf. *ante*, p. 43) Moreover the German play appears to be founded on Shakespeare's piece, not on a 'Pre-Titus'.[3]

3) Ravenscroft's assertion in 1687,

> I have been told by some anciently conversant with the Stage, that it (Titus Andronicus) was not Originally his (Shakespeare's), but brought by a private Author to be Acted, and he only gave some Master-touches to one or two of the Principal Parts or Characters—

is altogether unreliable. (See Mr. Crawford's paper, of unequal merit, in the Shakespeare Jahrbuch XXXVI, 109f.)

4) It has been questioned whether Shakespeare was the sole author of the play. But Shakespeare must have had a large share, to say the least, in the composition of the work, as Meres (1598) mentions it among other undoubted plays of Shakespeare, and as it is included in the First Folio. My view is that it is the *tour de force* of his early career.[4]

[1] The First Part of Henry VI. is probably the work of more than one author.

[2] Th. Eichhoff (Der Weg zu Shakespeare, Halle, 1902) says all the *ne*'s were forged. This is not correct. See below, Addenda and Corrigenda.

[3] Mr. Fuller, in the Publications of the Modern Language Association, Baltimore, 1901, vol. XVI, no. 1, discusses the question, but arrives at different results.

[4] Shakespeare might therefore be said to have begun with a play of blood and furious revenge, and ended with a play (The Tempest) of what Goethe calls *reine Menschlichkeit.*

For the existence of a PRE-RICHARD II. there is no shred of evidence. On the day before Essex's ill-fated revolt an 'old' play on the subject of Richard II. was acted by the Lord Chamberlain's men in compliance with a requisition by the Earl. This, I take it, was Shakespeare's play. I publicly defended a 'thesis' to this effect on the occasion of my 'promotion'. I was, and am, thoroughly persuaded of the correctness of my arguments, which I have not at my fingers' ends at the present moment.

The hypothesis of a PRE-AS-YOU-LIKE-IT, which Dr. Furness puts forward,[1] seems to me absolutely superfluous.

Mr. Verity, in his 'Students' Shakespeare' edition of Macbeth (1902), has collected all the arguments in favour of the theory of a PRE-MACBETH. The arguments make no impression on me. First, Mr. Verity says 'a "ballad of Macdobeth" was entered on the Register 'of the Stationers' Company August 27, 1596; and the same entry 'records "the ballad entituled The Taming of a Shrew"'. On turning to Arber's Transcripts, I find no trace of the above entry, which was first cited by Collier.[2]—Again, Mr. Verity refers us to 'Kemp's Nine Days' Wonder', 1600, where a penny Poet is mentioned, "whose first "making was the miserable stolne story of Macdoel, or Macdobeth, or "Macsomewhat, for I am sure a Mac it was", which, of course, is an allusion to the famous names *Do-wel* and *Do-bet* in 'Piers the Plowman'.

Again, the play called "Malcolm King of Scottes" which Mr. Verity mentions, and which we find entered in Henslowe's Diary, April 27, 1602, proves nothing. Nor can Dr. Gwinne's dialogue of about thirty lines, recited before the King at the gate of St. John's College in 1605, be looked upon as the germ of Shakespeare's play. It is only symptomatic of the time and the new interests evoked by the accession of a Scotch King. There is nothing to prove that Shakespeare's Macbeth is later in date than 1605.

I wish my reader to get rid, once for all, of the fiction of a hypothetical PRE-TEMPEST, which is imagined to have had much in common with the *Schöne Sidea*, a play by Jacob Ayrer, who died in 1605. As a matter of fact, there is not the slightest connexion and only the faintest possible resemblance between Shakespeare's Tempest

[1] New Var. Ed., vol. VIII, pp. 308 f.

[2] The entry *may* be in the Registers (cf. Fleay, 'Shakesp.', p. 241). Perhaps there is a reference to it in Kemp's Nine Days' Wonder. But Macdobeth is not=Macbeth.

and Ayrer's play. The plot of the former is so simple that there is no necessity for assuming either a narrative or a play for its source.

SOME SUGGESTED SOURCES.

Professor Koeppel[1] points out that a scene similar to that at the close of The Winter's Tale, where Hermione is shown as a statue, occurs in *The History of the Tryall of Chevalry,* Act IV, acted by the Derby's men (*not* Shakespeare's troupe, cf. Fleay, Stage, p. 154) and printed in 1605. A statue is also brought to life in Lyly's *Woman in the Moon.*

Mr. Boodle[2] draws attention to some resemblances—but not very striking—between Cymbeline and *The Rare Triumphs of Love and Fortune,* printed in 1589.

A Warning for Fair Women, acted by the lord Chamberlain's men shortly before its publication in 1599 has a passage very similar to one relating to Caesar's wounds in 'Julius Caesar' III, II.. If Mr. MacMillan[3] is correct in assigning to Shakespeare's play the date 1599, the passage in *A Warning* may be looked upon as containing an echo from Julius Caesar.

The Latin lesson in the Taming of the Shrew may have been suggested by a similar scene in *The Three Lords and Three Ladies of London* (printed 1590) by Robert Wilson, who was apparently a member of Shakespeare's company.

Resemblances between Shakespeare's plays and other dramas connected with Shakespeare's Company before 1611—of which Mr. Fleay has drawn up a list in his Life and Work of Shakespeare, 1886, p. 361f.—will no doubt be easily discovered.

THE 'PSEUDO-SHAKESPEAREAN' PLAYS

present a curious problem. Some of them belonging to Shakespeare's company may have received slight touches by the great poet. On the other hand, he must have been influenced by the plays in the performance of which he took part. I have not worked much at these plays, and have to waive a detailed discussion here.[4]

[1] Archiv für neuere Sprachen, 97, p. 329f.

[2] Notes and Queries, Nov. 19, 1887; and Shakesp. Jahrb. XXXIII, 344.

[3] The Arden Shakespeare edition of 'Caesar', 1902.

[4] Cf. Shakespeare Jahrbuch XXVII, 135f.; and Warnke and Proescholdt's edition.

MYSTERIES AND MORALITIES.

Shakespeare's acquaintance with these early forms of the English drama is testified by numerous allusions.

We have a reference to the Whitsun plays in The Two Gentlemen of Verona, IV, iv, 162f.:

> *Silvia.* How tall was she?
> *Julia.* About my stature; for at Pentecost,
> When all our pageants of delight were play'd,
> Our youth got me to play the woman's part,
> And I was trimm'd in Madam Julia's gown, etc.

Probably Shakespeare had seen performances of miracle-plays at Coventry, the home of pageants in the midland counties. Shakespeare's sketch of artisan players in A Midsummer Night's Dream contain probable reminiscences of his early years. It has been surmised that he witnessed some of the festivities at the Kenilworth Castle in 1575, when Leicester entertained the Queen with lavish magnificence with a view of gaining her hand at last,—but

> the imperial votaress passed on,
> In maiden meditation, fancy free.
> (cf. Mids. N. Dr., II, i, 148f.)

Herod and Termagant, furious and ranting characters of the old Miracles, are alluded to in the following passages:—

> O'erdoing Termagant; it outherods Herod
> (Hamlet, III, ii, 16);

> Herod of Jewry dare not look upon you
> But when you are well-pleas'd
> (Ant. and Cleop., III, iii, 3);

> What a Herod of Jewry is this?
> (Merry Wiv. II, i, 20).

We also find characters of the Old Morality alluded to, especially the clown, known as the Vice[1]:—

> Like to the old Vice . . . .
> Who, with dagger of lath,
> In his rage and his wrath,
> Cries, ah, ha! to the devil.
> (Twelfth Night, IV, ii, 130—141).

---

[1] Ed. Eckhardt is the author of Die Lustige Person im alten Drama. Berlin, 1902.

And now is this Vice's dagger become a squire . . . . .
(2. Henry IV., Act III, II, 343).

In 1. Henry IV., Act II, IV, 499, Prince Henry calls Falstaff

that reverend vice, that grey iniquity, that father ruffian, that vanity in years.

Vice, Iniquity, and Vanity occur frequently as characters in the Moralities, as also Justice—

Justice or Iniquity
(Meas. f. Meas., II, I, 181).

Again, Richard III., Act III, I, 82:

Thus, like the formal vice, Iniquity,
I moralize two meanings in one word.

Compare also Hamlet, III, IV, 98:

A vice of kings—

and Henry V., Act IV, IV, 75, where the Boy says of Pistol:

Bardolph and Nym had ten times more valour than this *roaring devil i' the old play,* that every one may pare his nails with the wooden dagger.

MASKS.

Masks may be roughly divided into three kinds. A mask in its simplest form is a mere disguise or masquerade without recitals. Under this head I would include masked dancing (such as we have in Much Ado, II, I; in Romeo, I, IV and V; Henry VIII., Act I, IV; and perhaps in Mids. N. Dr., V, 368—Bergomasc), and masked procession (as in Merchant of Ven., II, VI).

The second sort are masks joined with recitals and dancing,—such as we find in Love's Labour's Lost, Act V, II, 157 ff. Puck and the other fairies give a very masklike performance at the close of Midsummer Night's Dream. The maskers were ushered in by a herald or presenter, who introduced them to the audience by way of a prologue. In Love's Labour's Lost, V, II, Moth acts the herald of the Muscovite mask (cf. v. 97). The customary dance is, however, refused (v. 212). In 'Romeo' the part of a herald (who might have represented Cupid or the like, Act I, IV, *ad init.*) is purposely omitted.

More important is the third class of masks,—the Court-Masks, which found great favour in the eyes of King James and 'Bel-Anna', and became also popular on the public stage. They present a more developed and complicated form of the second sort[1]. We find them in-

[1] The first real 'court-masque' preserved to us seems to be that presented before the Queen in 1594. It is printed in Nichols's Progresses of Q. Elizabeth, III, 309. Is it authentic?

troduced in many of the later dramas of the time. In Shakespeare's Tempest a court-mask graces the union of Ferdinand and Miranda.[1] In the Archiv für neuere Sprachen CVII, p. 178, I had drawn attention to the resemblances between it and Beaumont's 'Masque of the Inner-Temple and Gray's Inn' (1613) and made the suggestion that it was written under the influence of the latter, which may have been originally played in place of the present mask. Professor Thorndike[2] points out that the mask of the Satyrs in The Winter's Tale, IV, IV, was taken over from Jonson's Masque of Oberon, acted on Jan. 1, 1611 before the King. Hence the remark: 'One three of them [the Satyrs] . . . 'danced before the king' (l. 345). The mask in Timon, I, II, was also suggested by the court-masks.

Lastly, the whole play of The Tempest with its characters of Ariel, Caliban (an antimask figure), and 'spirits in shape of dogs and hounds' (IV, I *end*), with its harpy-scene, and the entertainment (already referred to) by Iris,[3] Ceres, Juno, etc., has plainly experienced the influence of the court-masks, which, we may say in general, are responsible for much of the stage pageantry of Shakespeare's latest plays.

On popular performances, such as the Morris Dance, The Nine Worthies, and the like, referred to by Shakespeare, compare Dyer's Folk-Lore, and E. K. Chambers, The Mediaeval Stage, 1903.

[1] Why does Prospero warn Ferdinand not to break the virgin-knot? (IV, I, 15, 51 f.). The answer sounds frivolous, but I believe it is the only correct one. Prospero has to send the couple into the cell *all by themselves*, where they are afterwards discovered playing chess most innocently. That is why Ferdinand swears:

the *murkiest den*,
The most opportune place, the strong'st suggestion
Our worser genius can, shall never melt
Mine honour into lust.

[2] Cf. The Influence of Beaumont and Fletcher on Shakespeare, 1901:—Beaumont's same mask presented on 20 Febr. 1613 is also imitated in The Two Noble Kinsmen. See New Shakspere Soc., Ser. II, 15, part II, p. 53* or Thorndike, *ut sup.*, 44.

[3] Who plays the part of Iris? The answer is very clear: Ariel. Act IV, I, vers 49, Prospero had said to Ariel: "Do not approach Till thou dost hear me call". And shortly after he calls him: "Well. Now come, my Ariel!" Thereupon Iris enters.—Vers 167, "when I *presented* Ceres" does not therefore mean 'when I *acted* Ceres', as Schmidt says, but 'when I *heralded* or *introduced* Ceres'.

CHAPTER 5.

POPULAR LITERATURE.

INTRODUCTORY.

For contents of this chapter compare the general Table of Contents.

Dispersed through Shakespeare's text we find a considerable number of fragments and scraps of old ballads and songs, of which no further traces remain. Examples will occur to every reader. I have not collected these. It is to be hoped that a list of such of them, as have the air of a quotation from some lost poem, will be given in a later volume.[1] It is generally impossible to say with absolute certainty whether a song only known to us through Shakespeare's plays is to be attributed to the poet or not. Such dubious cases I pass over without notice.

Of the ballads and songs referred to by our poet I mention only such as have been preserved to us, or of which there are manifest traces outside of Shakespeare's works,—traces tending to show their contemporary or pre-Shakespearean existence, their vogue, etc. I have refrained from quoting in full or in part the originals. These poems will be published in collected form in a later volume. Such *portions* as are specially alluded to, or quoted, by Shakespeare will be italicized. Meanwhile it was necessary to give some bibliographical information.

In Shakespeare's days the minstrels had not yet died out. It was only in 1597 that a statute was passed by which minstrels wandering abroad were included among rogues, vagabonds, and sturdy beggars, and were adjudged to be punished as such. In 1589 Puttenham writes the following passage, which, though repeatedly quoted by others before, deserves a place here: 'The over busie and too speedy returne 'of one maner of tune, doth too much annoy and as it were glut the

[1] Compare my Preface *ante*.

'eare, unlesse it be in small and popular Musickes song by these '*Cantabanqui* upon benches and barrels heads where they have none 'other audience then boys or countrey fellowes that passe by them in 'the streete, or else by blind harpers or such like taverne minstrels 'that give a fit of mirth for a groat, and their matters being for the 'most part stories of old time, as the tale of Sir *Topas*, the reportes 'of *Bevis* of *Southampton*, *Guy* of *Warwicke*, *Adam Bell*, and *Clymme* 'of the *Clough*, and such other old Romances or historicall rimes, made 'purposely for recreation of the common people at Christmasse diners 'and brideales and in tavernes and alehouses, and such other places 'of base resort.'[1] (The Arte of English Poesie, Arber's Reprint, p. 96; comp., too, p. 57 ib.). I suppose it is unnecessary to add, that Sidney's 'Arcadia', Montemayor's 'Diana', etc., though also called Romances, are works of an entirely different character from those treated in the present chapter, and are noticed elsewhere.

Ballads fall into two classes: the Folk-ballads and the Art-ballads. Of the former the Robin Hood ballads seem to have been favourites with Shakespeare. I am sorry to say that there is no trace in his works of 'the old song of Percy and Douglas' which moved Sir Philip Sidney's heart 'more than with a trumpet' though 'sung but by some 'blind crowder'.[2] With the Popular ballads the Art-ballads contrast rather unfavourably. The chief ballad-makers known to us were Elderton and Deloney. But the rhymesters and cantabanks who balladized every trifling event or topic were legion, and their stall-ballads were sung and sold in every street. Falstaff refers to this practice in 1. Henry IV., Act II, II, 48:—

> An I have not ballads made on you all, and sung to filthy tunes, let a cup of sack be my poison.

And in a still more interesting passage in 2. Henry IV., Act IV, III, 49f.:

> here be [Colevile] is, and here I yield him and I beseech your grace, let it be booked with the rest of this day's deeds; or, by the Lord, I will have it in a particular ballad else, with mine own picture on the top on't, Colevile kissing my foot.

General notices of ballads are fairly numerous in Shakespeare's plays. The passages referring to them in The Winter's Tale (Act IV,

[1] Compare Lucr. 817—818:—"*Feast-finding minstrels*, tuning my defame,
"Will tie the hearers to attend each line,
"How Tarquin wronged me, I Collatine."
and Love's L. L., V, II, 405:—"Nor woo in rhyme, like a blind harper's song".

[2] Defence of Poesy.

sc. IV) are too well known to require a special mention. No doubt Mopsa echoed the opinion of many folks of those days, when she declared: "I love a ballad in print o' life" (l. 263). Evidently Shakespeare thought otherwise as regards the majority of the street-ballads.

Shakespeare, lover of music that he was, has numerous allusions to songs and tunes. 'During the long reign of Elizabeth, music seems 'to have been in universal cultivation, as well as in universal esteem'.[1] 'If ever a country deserved to be called "musical", that country was 'England, in the 16th and 17th centuries'.[2]

As for the dance tunes, it ought to be borne in mind that 'All 'the ancient dances were originally sung . . . . . In Elizabeth's reign 'some dances were sung, and others were simply played'.[3] Most of the popular dance tunes were set to words and sung.

Chappell's work on Old English Popular Music as well as his and the Rev. J. W. Ebsworth's publications for the Ballad Society have been of invaluable aid to me. I have occasionally had to refer to the earlier edition of the 'Popular Music' (1855—59), as it contains important information, expunged, for what reason I do not know, from the latest edition, 1893, by Mr. Wooldridge.

The chapter concludes with Popular Tales and Light Literature. The list of popular tales might be further extended by including such legends as that of Patient Grissel (mentioned in The Shrew, II, I, 297), and the like. Of Jest-books a collection consisting of three volumes has been printed by W. Carew Hazlitt in 1864 under the title 'Shakespeare Jest-Books'. It was the custom of writers of the old jest-books to attribute them to some well known personage in order 'to throw 'a halo of popularity round their facetious lucubrations' and to secure a more ready market. This same charge must be laid at the door of W. C. Hazlitt, whose 'Shakespeare Jest-Books', save one or two, have nothing whatever to do with Shakespeare.

THE HEROES OF ROMANCE.

Of all the heroes of Romance the three most famous in England were Arthur, Guy of Warwick, and Sir Bevis, all three of whom are referred to by our immortal poet. With

[1] Chappell, Old Engl. Pop. Music, 1893, vol. I, 59.
[2] Naylor, Shakespeare and Music, 1896, p. 20.
[3] Naylor, pp. 113—4.

THE ARTHURIAN LEGENDS

(whether in the form of prose, metrical romance, or ballad, from books, broadsides, or from the mouth of 'Cantabanqui', or minstrels, etc.) Shakespeare is familiar, as might, of course, be expected. "*Queen Guinover of Britain*", Arthur's wife, is mentioned in Love's Lab. Lost (IV, I, 125); and *Merlin*, the enchanter and prophet of the Arthurian romance, in 1. Henry IV. (Act III, I, 150): "*the dreamer Merlin and* "*his prophecies*" [referred to by Hall in his chronicle, Bosw.-Stone, p. 139[2]], and in King Lear, III, II, 95: *Fool* . . . . . "*This prophecy* "*Merlin shall make;* for I live before his time".—*Camelot*, where Arthur's residence was, is mentioned in Lear, II, II, 90. In 1. Henry VI. (Act III, II) 'Bedford, who has been "brought in sick in a chair" '(l. 40), determines to "sit before the walls of Rouen" (l. 91), awaiting 'the issue of an attempt to regain the city, for he has "read"'[1]

> That *stout Pendragon in his litter sick*
> *Came to the field and vanquished his foes:*
> Methinks I should revive the soldiers' hearts,
> Because I ever found them as myself.
> (ll. 95—98.)

Shakespeare certainly did not read this in Holinshed, as Holinshed attributes this deed to the *brother* of Pendragon, Arthur's father. But Malory, in his Le Morte Darthur, chapter IV, gives an account which agrees with that known to Shakespeare. [Geoffrey of Monmouth, too, and Hardyng ascribe the heroic deed to Pendragon.]

In 2. Henry IV. (Act III, II, 298) Shallow says:—

> I remember at Mile-end Green, when I lay at Clement's Inn,—I was then *Sir Dagonet in Arthur's show.*

Arthur's show was an exhibition of archery at London, in which also Sir Dagonet, King Arthur's fool, figured.

What Arthur the Hostess, in Henry V., Act II, III, 10,—"he's in Arthur's bosom"—is thinking of, it would be unsafe to say.[2]

In 2. Henry IV. we find a quotation from THE NOBLE ACTS OF ARTHUR, OF THE ROUND TABLE, AND OF LANCELOT DU LAKE. TO THE TUNE OF "FLYING FAME". Such is the title of the ballad given in *The Garland of Good-Will* 'written by T[homas] D[eloney]' (who died *circa* 1600) which is dated by Henry Dixon *circa* 1586, but of which we possess only later copies. A reprint of the 'Garland' will be found

[1] Boswell-Stone, Shakspere's Holinshed, p. 226.

[2] Of course, she means *Abraham's* bosom.

in the Percy Society, vol. XXX. The above ballad is on p. 38, and its first stanza runs thus:

When Arthur first in court began,
And was approvèd king,
By force of arms great victories won,
And conquests home did bring; etc.

The ballad is nothing more than a rhymed version of certain chapters of Sir Thomas Malory's 'Le Morte Darthur'. 'Falstaff', as John W. Hales puts it, 'quotes the first line except the last word, and after 'a brief interruption the second, which he makes "And was a worthy 'king", in the 2nd part of "King Henry IV.", Act II, sc. IV, 36. It 'is quoted also, as Mr. Chappell [I, 92] mentions, in Marston's "Mal-'content", and Beaumont and Fletcher's "Little French Lawyer".' (Repr. of Percy's Fol. MS., vol. I, p. 84). The ballad, I may add, is also quoted in Th. Heywood's 'Rape of Lucrece', first printed in 1609,

When Tarquin first in Court began,
And was approved King:
Some men for sudden joy gan weep,
But I for sorrow sing.

With the last two lines compare Lear, I, IV, 191—2. As to the tune to which Falstaff must have hummed the words compare Chappell, *ut sup.*

I ought to mention, that Falstaff's words in 2. Henry IV. (Act II, IV, 53): "Your brooches, pearls and, ouches" are looked upon by the editors as a quotation from The *modern version* of THE BOY AND THE MANTLE, beginning 'In Carleile dwelt King Arthur' (printed by Percy), the 11th line of which is: 'With brooches, rings, and owches'. I know nothing definite about the date of composition of this poem 'as revised and altered by a modern hand'. I rather suspect the modern hand was Percy's own. In the older version 'ouches' is not given.[1]

[1] Falstaff's words must be taken in close connexion with the preceding passage in the text:

Falst. .... "we catch of you, Doll" ....
Doll: "Yea, joy, our chains and our jewels."
Falst. "Your brooches, pearls, and ouches: for to serve bravely is to come "halting off, you know."

Brooches, pearls, and ouches, of course, refer to something very different from gems. Carbuncle is similarly frequently used by us in the sense of tumour, pimple. Falstaff is *not* quoting from any ballad. (Cf. Comedy of Err., III, II, 137—8, quoted *infra*, p. 232.)

GUY OF WARWICK.

For bibliographical matter relating to Sir Guy I must refer the curious reader to Zupitza's edition, Early Engl. Text Soc., and to 'Percy's Folio MS.' II, 509f.[1] It will be sufficient here to say that the romance existed in print in the 16th century. It was also condensed into a ballad *circa* 1592. And numerous allusions in Elizabethan literature testify to its popularity and persistence. Shakespeare has two: In King John, I, I, 225, the Bastard derisively names his weak half-brother "*Colbrand the giant*, that same mighty man". This Colbrand, the champion of the Danes, who is with difficulty defeated by Guy, is again referred to in Henry VIII., Act V, IV, 22: "I am not "Samson, nor *Sir Guy, nor Colbrand*, To mow 'em down before me."

SIR BEVIS OF HAMPTOUN.

was a no less popular hero of romance in the sixteenth century. For particulars concerning old MSS. and editions, see Furnivall's lucid Introduction to Robert Laneham's Letter, and Kölbing's edition, Early Engl. Text Soc. The story is sketched by Ellis, in his Early Engl. Metr. Rom. A summary is also given by Kölbing.

In Henry VIII. (Act I, I, 33f.) we have the following allusion:

When these suns [the two Kings]—
For so they phrase 'em—by their heralds challenged
The noble spirits to arms, they did perform
Beyond thought's compass; that *former fabulous story*,[2]
Being now seen possible enough, got credit,
That *Bevis* was believed.

In King Lear (III, IV, 144) we find a couplet, a little altered, from this old Romance in the mouth of Edgar:

But mice and rats, and such small deer,
Have been Tom's food for seven long year.

The original verses are part of the description of Bevis's adventures in Damascus, where he was kept in prison seven years. 'Some wheat-

[1] I need scarcely add, that the stories of 'Guy', 'Bevis', etc. are given in a modern dress by Ellis in his 'Specimens of Early English Metrical Romances' (Bohn's Libr.).—French, in his 'Shakespeareana Genealogica', pp. 580 & 431 ff., asserts that the great dramatist is a lineal descendant of Sir Guy's. I mention this as a mere curiosity. "Fiddle-faddle", says J. W. E.

[2] former fabulous story=the story of old times, hitherto thought fabulous.

'bran was daily let down into the dungeon for his support: but neither 'meat nor corn was allowed to him; and

Rattes and myce and suche smal dere
Was his mete that seven yere'.[1]

Besides these two references in Shakespeare's text, we find a third in The First Part of the Contention, which is either an imperfect version of 2. Henry VI., or, as some maintain, the original of it:

have at you Peter with downright blowes, *as Bevys of South-hampton fell upon Askapart.*

The words in italics are wanting in 2. Henry VI. (cf. Act II, III, 92).[2] Ascapart was a formidable giant subdued by Bevis, whose page he then became.

SIR EGLAMOUR.

'Shakespeare may possibly have had this hero in his mind when 'he calls one of his characters by his name in the Two Gentlemen of 'Verona: "What think'st thou of the fair Sir Eglamour"'[3] (Act I, II, 9). This knight "valiant, wise, remorseful, well accomplish'd" becomes the agent for Silvia in her escape in a later part of the drama. The Romance was 'printed at Edinburgh in 1508 by Walter Chapman, 'and subsequently at London by Copland and Walley'. It is entered on the Stat. Reg., Jan. 15, 1582.

THE SQUIRE OF LOW DEGREE.

This is the title of a most popular romance, which begins thus: "It was a squyer of lowe degrè".—It is alluded to by Fluellen:

"You called me yesterday mountain-squire; but I will make you to-day "*a squire of low degree.*"

(Henry V., Act V, sc. I, 36). A reprint will be found in Hazlitt's 'Early Popular Poetry', vol. II.[4]

[1] Ellis, *ut sup.*, p. 256; Kölbing, p. 75, var. lectio.

[2] They are wanting in the Folio. Perhaps they ought to be inserted into Shakespeare's text.

[3] These are J. O. Halliwell's words (Percy's MS., II, 338). He adds 'The name, 'however, appears to have passed into a kind of proverb. So in Dekker's *Satiro-'mastix:* "Adieu, Sir Eglamore! adieu, lute-string"', etc. I find nothing in this passage, which points to a proverbial use of the name. The words referred to by Halliwell are spoken by Tucca, who alludes to all manner of persons and things under the sun *à la* Pistol.

[4] 'The squyre of Low deggre' I find entered in the Stat. Reg., 10. June 1560. (Arber, I, 128)

SIR TOPAS

deserves a passing notice and I prefer to mention him here. He is made the hero of a burlesque ballad by Chaucer in his Canterbury Tales. 'The tale of Sir *Topas*' is mentioned by Puttenham (cf. *ante*, p. 156) as forming part of the stock of the Cantabanqui and Minstrels. The name, Sir Topas, is assumed by the clown in Twelfth Night (IV, II). But 'Sir Tophas' is also in Lyly's Endymion.

CHARLEMAGNE ROMANCES.

Enormous though the popularity was which these romances enjoyed in the middle ages, the English versions never took very deep root. It is therefore not suprizing to find but few and faint traces of the Charlemagne cycle in Shakespeare's works.

In 1. Henry VI., Act I, II, 29—31, Alençon is made to refer to the two most famous paladins of Charles the Great:

> Froissart, a countryman of ours, records,
> England all *Olivers and Rowlands* bred
> During the time Edward the Third did reign.[1]

Of the Charlemagne Romances, there is only one with which, we can say, Shakespeare is in some way connected, whether directly or indirectly. This is 'Huon of Bordeaux', which was translated from the French original by Lord Berners,[2] about 1530, who, be it remembered, was also the translator of Froissart's Chronicle. Of the popularity of Huon of Bordeaux in England there is no lack of evidence. We know, too, from some entries in Henslowe's Diary, that the romance had been dramatized and produced on the London stage in 1593 and 1594. From these and other considerations we see that Oberon, who, in the old romance, plays almost as important a *rôle* as Huon himself, had been naturalized in England long before Shakespeare classicized him for ever, in his Mids. Night's Dream.

Apart from the name and the sovereignty in fairy land, Shakespeare's Oberon has in common with his medieval predecessor the circumstance that his kingdom is situated in the far East. Compare, *e. g.*, Mids. N. Dr., II, I, 68, where Titania asks Oberon:

> Why art thou here,
> Come from the farthest steppe of India?

[1] The particular passage in Froissart, if any, alluded to here, has not been identified yet.

[2] Edited, with an introduction, by Sidney Lee, Early Engl. Text Soc., XL.

It is possible, too, that we have an allusion to 'Huon of Bordeaux' in Much Ado About Nothing, II, I, 271ff., where Benedick declares he would rather fetch a toothpick from the furthest inch of Asia, bring the length of Prester John's foot, fetch a hair off the great Cham's beard, than hold three words' conference with Beatrice. Here there may be a reference to the grotesque task imposed upon Huon who was to go to Babylon and, among other things, to rob the 'Admiral (!) Gaudis' of a handful of hair from his beard and of four of his largest teeth.[1]

FOLK-BALLADS.

ROBIN HOOD.

William Shakespeare's fondness for Robin Hood, the people's darling, celebrated in so many ballads, is evinced by some striking allusions in his dramas. First, in As You Like It (I, I, 119) we have a passage which might be set as a fit motto before this exquisite pastoral drama:

Oliver: Where will the old duke live?

Charles: They say he is already in the forest of Arden, and *a many merry men* with him; and *there they live like the old Robin Hood of England:* they say many young gentlemen flock to him every day, and fleet the time carelessly, as they did in the golden world.[2]

Under the *greenwood tree* Robin Hood and his merry men of the ballads have their abode. Compare As You Like It, II, v, 1:

Under the greenwood tree[3]
Who loves to lie with me, etc.

Compare also 'My Robin is to the Greenwood gone' (see *post*, p. 178).

The two most prominent companions of Robin Hood were Little John and Scarlet, referred to by Shakespeare in 'The Merry Wives', I, I, 177, where Falstaff addresses the red-faced Bardolph as "Scarlet and "John", and in 2. Henry IV., Act V, III, 107, where Silence sings, "And "Robin Hood, Scarlet, and John". This is a line occurring in THE JOLLY PINDER OF WAKEFIELD (Child III, 131) st. 3:

[1] Of 'The King of *Fairies*', a play referred to by Nash (1589) and Greene (1592), nothing further is known (cf. Fleay, 'Drama', II, 279, 283). I cannot think that Greene's Oberon (in James IV.) is the father of Shakespeare's Oberon.

[2] The German translation by Schlegel is wretched: 'da leben sie wie *Zigeunervolk* . . . . und *versaufen* sorglos die Zeit wie im goldnen Alter'.

[3] The whole line occurs frequently, as a standing phrase, in the R. H. ballads. Compare, e. g., Child III, p. 71, st. 310, 312; p. 72, st. 328; 335; p. 74, st. 377; p. 97, st. 2; 98, st. 23; 113, st. 83; 115a, etc.

All this beheard three witty young men,
'Twas Robin Hood, Scarlet, and John.

The ballad was well known in Shakespeare's days, being quoted from in three contemporary dramas, and mentioned as early as 155⅞ in the Stationers' Registers.[1] It is therefore extremely probable that this is the ballad which Shakespeare is quoting.[2]

In The Two Gentlemen of Verona, Acts IV & V, we find a band of civil robbers introduced, who "detest such vile base practices" as "outrages on silly women or poor passengers". Of their resemblance to Robin Hood's outlawed company Shakespeare is so fully aware that he puts the following expression into the mouth of one of them: "By the "bare scalp of Robin Hood's fat friar." This Friar, however, known as Tuck, does no more belong to the popular Robin Hood *ballads* than Maid Marian (mentioned in 1. Henry IV, Act, III, III, 129) both of whom owe their association with Robin Hood primarily to the May-games and morris dance. 'In the truly popular ballads Friar Tuck is never 'heard of,. and in only two even of the broadsides, Robin Hood and 'Queen Katherine and Robin Hood's Golden Prize, is he so much as 'named; in both no more than named, and in both in conjunction 'with Maid Marian,' 'who appears elsewhere only in a late and entirely 'insignificant ballad'. (see Child, III, 122 & 43f.)[3]

A Note.

It is supposed that the following passage in Much Ado (I, I, 259),

Benedick: If I do, hang me in a bottle like a cat and shoot at me; and he that hits me, let him be clapped on the shoulder, and called Adam,—

contains an allusion to the famous archer *Adam Bell*, celebrated in a wellknown ballad: 'Adam Bell, Clim of the Clough, and William of Cloudesly' (Child, III, 14). But this is by no means self-evident.

[1] As to the old tune, comp. Chappell (old ed. 393).

[2] It must be admitted, however, that the names Robin Hood, Scarlet, and John are to be found together, in this order, in a later version of R. H. and Queen Katherine 1,2 (Child, III, 202); Compare, too, Child, III, 147, st. 1; III, 132 b, c; III, 171, st. 1. A passage:

"No man may compare with Robin Hood,
"With Robin Hood's Slathbatch and John a"—

occurs in a poem in the prose 'History of George a Green' (see Thoms, Early Engl. Prose Rom., 1858, II, 189.—Slathbatch=Scarlet). And who knows whether some lost ballad does not contain the same words quoted by Silence.

[3] A Robin Hood *motif* is introduced into 1. Henry VI., Act II, III, where Talbot, imprisoned by the Countess of Auvergne, summons his companions by blowing his horn

NARRATIVE ART-BALLADS.

A SONG OF A BEGGAR AND A KING.

(KING COPHETUA AND THE BEGGAR-MAID.)

This ballad is preserved in 'A Crowne-Garland of Goulden Roses' (1st. ed., 1612) by Richard Johnson, reprinted by the Percy Society, vol. VI. It was repeated by Percy in his Reliques.[1] The ballad of King Cophetua is referred to by Shakespeare on five different occasions:

1) In Love's Lab. Lost, I, II, Armado, who is "in love with a "base wench", asks Moth: "what great men have been in love?" Moth reminds him of Hercules and Samson. Not content with these authorities he says (l. 114):

> Is there not *a ballad*, boy, *of the King and the Beggar*?
>
> *Moth:* The world was very guilty of such a ballad some three ages since: but I think now 'tis not to be found; or, if it were, it would neither serve for the writing nor the tune.
>
> *Arm.* I will have that subject newly writ o'er, that I may example my digression by some mighty precedent.

2) To the same ballad Armado's long letter (Love's L. L., IV, I, 60—88) is full of allusions. The beggar-maid is here called Zenelophon. In the ballad, *P*enelophon.

3) Richard II., Act V, III, 77:—

> *The Duchess* (within):
> Speak with me, pity me, open the door:
> A beggar begs that never begg'd before.
> *Bolingbroke, the King:* Our scene is alter'd from a serious thing,
> And now changed to '*The Beggar and the King*'.

4) The opening words of the second stanza of the ballad are particularly alluded to in 'Romeo' (II, I, 11f.):

> Speak to my gossip Venus one fair word,
> One nick-name for her pur*blind son* and heir,
> Young Adam *Cupid, he that shot so trim,*
> *When King Cophetua loved the beggar-maid!*

5) A fifth allusion is in 2. Henry IV., Act V, III, 105—6, where Falstaff affectedly says to Pistol:

> O base Assyrian knight, what is thy news?
> Let King Cophetua know the truth thereof.

---

[1] Also Ballad Soc., Roxb. Ball., vol. VI, pp. 659, 661, both versions. Tennyson, it is almost impertinent to mention this, has written a poem on the same subject.

Malone thinks that the words, "The king's a beggar" in All's Well, V, III, 335, Epilogue, contain 'some allusion to the old tale of The King and the Beggar'. The allusion is, however, not very clear.[1]

THE CONSTANCY OF SUSANNA.

The first line: 'There dwelt a man in Babylon' and the burden, 'Lady, lady', of this ballad are quoted by Sir Toby in Twelfth Night (II, III, 84). It was entered on the Stationers' Registers in 1562/3 (Arber, I, 210) under the title: 'the godly and constante wyse [=wife] Susanna.' But we only possess a later copy of it of the reign of James I., to be found in the Roxburghe Collection, I, 60[2], with the following title: 'An excellent Ballad / Intituled: / The Constancy of Susanna. / To an excellent new tune.'

The ballad was attributed by Chappell to William Elderton chiefly (as it appears) on the score of identity in the measure and burden with a production of W. E.'s, entitled "The Panges of Love and Lovers Fittes", 1559-60.[3] There are, however, four more productions of this sort, three anonymous, and one signed R. M.[4]

In 'Merry Wives', III, I, 24, Evans, the Welsh parson, sings the first line of the metrical version of the 137th Psalm: "When as I "[*loco* we] sat in Pabylon." The First Quarto here gives: "There "dwelt a man in Babylon". But the First Quarto is not a reliable authority and the other reading is more in character.

It has been supposed, that the words, Mercutio addresses the Nurse with, in 'Romeo', II, IV, 151: "Farewell, ancient lady; farewell, lady, "lady, lady," contain an allusion to the burden referred to above. But this is very doubtful.[5]

[1] Ben Jonson refers to the ballad in Ev. Man in his Hum. III, II: 'rich as 'King Cophetua'. Instead of 'Cophetua', however, of the Folio edition 1616, the Quarto (1601) reads 'Golias'.

[2] Ballad Soc., Roxb. Ballads, vol. I, p. 190, where also a note on the date and on other copies of the ballad is given by Chappell.

[3] reprinted by the Percy Soc., vol. I, 'Old Ballads', p. 25.

[4] First, the anonymous fragment in A Handefull of pleasant delites (1584), Arber, Engl. Sch. Libr. 3. p. *25*. Then, the anonymous composition in 'The Trial of Treasure,' pr. 1567 (Dodsley, III, 292); Thirdly, 'A newe Ballade', signed R. M., beginning 'O dere Lady Elysabeth' ('belonging apparently to a rather early period of the queen's reign') reprinted in 'Ancient Ballads and Broadsides', Lilly, 1870, p. 30. A fourth example is found in Twenty-five Old Ballads and Songs, from MSS. in the possession of J. Payne Collier, 1869, p. 19.

[5] Tyrwhitt, Malone's Var. Ed., XI, p. 395 (Twelfth Night), compares Ben Jonson's Magnetic Lady', Act IV, III,—*Compass: "As true it is, lady, lady,* in the

JEPHTHAH JUDGE OF ISRAEL,

a ballad quoted by Hamlet (Act II, II, 422f.). A copy of it was given by Percy in his 'Reliques'. But a more perfect one is reprinted by the Ballad Society (Roxb. Ball., vol. VI, p. 685-6), by Halliwell (cf. Furness III, 174), and others. The supposition that 'a ballett intituled 'the songe of Jesphas Dowgther at his death,' entered on the Stat. Reg. 156$\frac{7}{8}$, is identical with our "pious chanson" is open to serious doubts.

REMARKS ON THE BALLADS OF TITUS ANDRONICUS, THE JEW OF VENICE, AND KING LEIR.

"Wenn die Könige bau'n haben die Kärrner zu tun."
Schiller.

1. Titus Andronicus's Complaint, reprinted, like the following two ballads, by Percy, is probably of later date than the play. The Stationers' Registers go some way towards proving this; for on the 6th of February 1594 'A Noble Roman Historye of Tytus Andronicus', and also 'the ballad thereof' were registered; but 'titus and ondronicus' was acted on January 23. of the same year.[1]

Both

King Leir and his Three Daughters,

and

Gernutus the Jew of Venice,

are now generally considered *not* to have contributed in any way towards Shakespeare's plays, treating the same subjects. They are probably later productions, owing their existence to the success of previous plays. (Comp. Furness, V, 402f.; VII, 288f.)

song". These words ought to be taken in connexion with those immediately preceding, spoken by Lady Loadstone: "I bid God give you joy, *if this be true*", as appears from the lines: "*If this be true, as true it was, Lady! lady!*" to be found in the '*Pangs*' (1st stanza) and in 'A newe Ballad' (st. 7), both referred to *ante*. Comp., too, Old Sh. Soc. 1844 p. XXXIV.—The commentators of Jonson have failed so far to give a satisfactory explanation of the passage.

[1] Compare my remarks *ante*, p. 149, and see below, Addenda and Corrigenda.

A Note.

Every reader of Percy's Reliques should bear in mind that *The Friar of Orders Grey* consists of fragments from Shakespeare and one from Fletcher strung together by Percy himself. Stanza 1[1,2] is from the Shrew, IV, I; 3, 5, 7 are, wholly or in part, from Hamlet, IV, V; 12, 13 from Fletcher's Queen of Corinth, III, II; 15 from Hamlet, as before; 17, 18 from Much Ado, II, III; one line of 22 from Lear III, IV. (Child, V, 201). If the reader remembers this, he will not commit the blunder of referring, as Alex. Schmidt did, to this poem as an authority for a textual reading. (Cp. Furness, V, 185.) Even the New Engl. Dict., our highest lexicographical authority, is guilty of this amazing *lapsus*, making certain lines sung by Ophelia belong to the Friar of Orders Gray! and quoting them from Percy's, not from Shakespeare's, text. (Cp. 'cockle' sb[2]. 7—p. 574.)

SONGS AND TUNES.

THE AGED LOVER RENOUNCETH LOVE,
beginning,
I LOATHE THAT I DID LOVE.

From this song by Thomas Lord Vaux, printed by Tottel in his Miscellany, 'Songes and Sonettes', 1557, etc. (see Arber's Reprint, p. 173), the grave-digger in Hamlet (Act V, I) sings three stanzas though much corrupted by him. The music is given by Chappell, I, 52, and by Rimbault, 'Musical Illustrations of Percy's Reliques', 1850, p. 48.[1]

[1] Malone thinks the Book of 'Songs and Sonnets' for which Slender sighs in The Merry Wives (I, I, 206) was Tottel's Miscellany. But the words 'songs and sonnets' I have found together rather frequently in Elizabethan Literature.
It is worth remarking here, that Goethe makes use of the clown's song in the Second Part of Faust:

Lemuren (mit *neckischen* Geberden grabend)

'Wie jung ich war und lebt' und liebt',
'Mich deucht, das war wohl süsse;
'Wo's fröhlich klang und lustig ging,
'Da rührten sich meine Füsse.
'Nun hat das tückische Alter mich
'Mit seiner Krücke getroffen;
'Ich stolpert' über Grabes Tür,
'Warum stand sie just offen!'

Note, that 'Krücke' is from the original. Instead of 'crutch' Shakespeare has 'clutch'.

BELL MY WIFE,
(TAKE THY OLD CLOAK ABOUT THEE),

beginning, 'This winter's weather it waxeth cold'. The earliest allusion we have to this song is to be found in Greene's Quip for an Upstart Courtier, 1592.[1] The earliest extant version of it is that contained in Percy's Folio Ms. (reprinted by Hales and Furnivall, vol. II, 322. A photolithographic impression of it is bound up with vol. I). Shakespeare quotes the whole last stanza but one of the old song in Othello, II, III, 92—99: "King Stephen was a worthy peer", etc. There is another evident allusion to this same stanza in The Tempest, IV, I, 221:

> *Trinc.* O *king Stephano!* O *peer!* O *worthy* Stephano! look what a *wardrobe* here is for thee!

And again l. 225:

> O *king Stephano!*

CALEN O CUSTURE ME.

This tune known by the above Irish name is alluded to by Pistol in Henry V., Act IV, IV, 4:

> Qualtitie *calmie custure me!*

Pistol is repeating the burden of an old song, 'Calen o Custure me', as Malone was the first to point out, quoting the following title from Clement Robinson's 'A Handefull of pleasant delites', 1584 (reprinted by the Spenser Soc., Issue No. 8, and by Arber, Engl. Sch. Libr., 3):—

> A Sonet of a Lover in the praise of his lady. To Calen o Custure me: sung at everie lines end.

The first line runs thus:

> When as I view your comly grace, *Ca.* &c.

Perhaps the song here referred to is identical with one that was licensed to J. Aldee on the 10th of March 1582, Arber, Transcr., II, 407:

> Tollerated to him twoe ballades whereof th[e] one [is] intituled *Callin o custure me* and th[e] other concerneth *the danger of Sailers* ....

Calino Casturame is one of the airs contained in the Fitzwilliam Virginal Book and reprinted by Chappell, I, 84. "Sir Robert Steward "(Grove's Dictionary of Music, art. Irish Music) says, 'It is evidently "an attempt to spell as pronounced the Irish phrase, "Colleen, oge "astore!"—young girl, my treasure!"' (see Clarend. Press ed. of

[1] Furness, VI, 131.—Regarding the tune see Rimbault, *ut supra*.

Henry V., p. 179). *Calino* is the tune of 'A pleasant Song made by a Souldier', date 1588, reprinted by the Ballad Soc., Roxb. Ballads vol. VI, p. 284.

CANST THOU NOT HIT IT?

is given as an old dance tune by Chappell (I, 249). In Love's Lab. Lost, IV, I, 127—130, Rosaline and Boyet recite or sing four lines which are a quotation from, or an imitation of, some part of the original song sung to that tune:

> *Boyet:* But she herself is hit lower: have I hit her now?
> *Ros.* Shall I come upon thee with *an old saying*, that was a man when King Pepin of France was a little boy, as touching the hit it?
> *Boyet.* So I may answer thee with one as *old*, that was a woman when Queen Guinover of Britain was a little wench, as touching the hit it.
> *Ros. Thou canst not hit it, hit it, hit it,*
> *Thou canst not hit it, my good man.*
> *Boyet. An I cannot, cannot, cannot,*
> *An I cannot, another can.*

THE CAREFUL LOVER COMPLAINETH,
AND THE HAPPY LOVER COUNSELLETH,[1]
beginning,
A [!] ROBYN JOLY ROBYN.

In his Twelfth Night (IV, II, 78—85) Shakespeare introduces the clown singing part of the two first stanzas of this song which has been recovered from an ancient MS. which seems to have been written in the reign of Henry VIII. (I am here quoting from Percy.)

COME O'ER THE BOURNE, BESSY,

is quoted from by Edgar in King Lear, III, VI, 27:

> Come o'er the bourn,[2] Bessy, to me.—

And what follows from the Fool:—

[1] The title according to Halliwell, Fol. ed., VII, 409.—Comp. Anglia XII, p. 241, and p. 272.

[2] *broome* is in the original text. 'Over the *broome*, Bessy' is the title of the tune in Univ. Lib. Camb. Lute MSS. (Chapp., I, 121). 'Broome' (=brook) is etymologically probably the same word as 'bourne', the change being due to metathesis. Comp. Arber, Transcr., I, 262: 4 Sept. 1564, A ballad licensed to Wm. Pickering, 'A saynge betwene the quene and Englonde Called comme over 'the *browne* Bessye to me.'

Her boat hath a leak,
And she must not speak
Why she dares not come over to thee—

no doubt belongs to the same old song, now lost, of which we possess two imitations (Chappell ,I, 121-2). Apparently the refrain of the *original* song is preserved to us in W. Wager's Morality 'The Longer thou livest the more Fool thou art' (*circa* 1560),[1] where Moros sings the 'foot' or burden:

Com over the Boorne, Besse,
My little pretie Besse,
Com over the Boorne, besse, to me.

The same burden (with the single change of 'thou' for 'my') is to be found in a moralisation reprinted in Chapell, *ut sup.*, where also the old music is given. The original song and tune must be assigned to the early years of the sixteenth century.

THE CROWE SITS UPON THE WALL,
PLEASE ONE AND PLEASE ALL.

'Please one and please all' is the beginning and the burden of the song by R. T., entered on the Stat. Reg. 1591-2. Only one copy is known to exist.[2] See note in Furness, XIII, p. 217-8. The heading of it is: 'A prettie newe Ballad, intytuled: The Crowe sits upon 'the wall, Please one and please all. To the tune of, Please one and 'please all.' The words: 'To the tune of, Please', etc. *may* imply the existence of an earlier tune (and song). The poem is alluded to in Twelfth Night, III, iv, 21f.:

Malv. Sad, lady! I could be sad: this does make some obstruction in the blood; this cross-gartering; but what of that? if it please the eye of one, it is with me *as the very true sonnet is, 'Please one, and please all'.*

FAREWELL, DEAR LOVE
(or Corydon's Farewell to Phillis).

This song, beginning: 'Farewel, dear love, since thou wilt needs 'be gon', first published, with the music, in Robert Jones's 'First Booke of Songes or Ayres', 1600,[3] is quoted from (with slight variations

[1] Reprinted in Shakesp. Jahrb. xxxvi, with an introd., by Prof. Brandl.

[2] Reprinted in Anc. Ball. and Broads., 1870, p. 255.

[3] The words and music of 'Farewell, dear love' are contained in the *First* Book of Songs which appeared in *1600.* The *Second* Book is dated 1601. (Br. Mus.; See also Davey's Hist. of Engl Music.)

to suit the occasion) by Sir Toby and the clown in Twelfth Night (II, III, 109—121). The words were repeated in 'The Golden Garland of Princely Delight', by Rich. Johnson (a 13th ed. 'with Additions' 1690; the 3rd 'enlarged', 1620), from which Percy reprinted the song in his Reliques. The music in Rob. Jones's work is reprinted by J. S. Smith in his 'Musica Antiqua', 1812, and by Dr. Rimbault in his 'Musical Illustrations of Bish. Percy's Reliques', 1850, see pp. 9 & 52. The first two stanzas of the original song are given by Halliwell-Phillipps, Outlines, II, 282, and the whole song in A. H. Bullen's 'More Lyrics from Elizabethan Songbooks', 1888, p. 33, and by Arber in his 'Anthologies'.

FIRE, FIRE.

In the 'Taming of the Shrew', IV, I, Grumio calls upon Curtis to make fire:

> A fire, good Curtis.
> *Curt.* Is my master and his wife coming, Grumio?
> *Gru.* O, ay, Curtis, ay: and therefore *fire, fire; cast on no water.*

This seems to be in allusion to a madrigal contained in Thomas Morley's well-known First Booke of Balletts which appeared in 1595.[1] This madrigal, consisting of two short stanzas, begins thus:

> Fyer, fyer, my hart!

and ends with:

> O cast cast water on alas and drench mee. Fa la la.

In illustration of the above passage in 'The Shrew' the celebrated Blackstone had quoted 'an old popular catch [=round] of three parts':

> Scotland burneth, Scotland burneth.
> Fire, fire;—fire, fire;
> Cast on some more water.

(Malone, Var. Ed., V, p. 457.)

FORTUNE MY FOE.

Shakespeare is supposed to allude to this then well-known song and tune (see Chappell, I, 76) in the following passage:

> I see what thou wert, if *Fortune thy foe* were not, Nature [being?] thy friend.

(Merry Wives, III, III, 69.) The first stanza of the old song is con-

[1] The words have been reprinted in 'Thomas Morley und die Englischen 'Madrigalisten der Shakespeare-Zeit', 1903, by Dr. Bolle, who drew my attention to the song.

tained in The Maid's Metamorphosis[1], a drama printed in 1600. The whole song is reprinted by Mr. Ebsworth in Ballad Soc., Bagford Ballads, p. 961. I ought to mention the fact, too, that the ballad of Dr. Faustus, entered on the Stat. Reg. 1588—9, and the ballad of Titus Andronicus (see above) are to the tune of Fortune my Foe.

THE GOD[S] OF LOVE,

the first four lines of which[2] are quoted by Benedick in Much Ado, V, II, 26—29,

The god of love
That sits above,
And knows me, and knows me,
How pitiful I deserve,—

is an old song (now no longer extant) by William Elderton.[3] 'The 'godes of Love' was licensed for printing in 1567/8. (Arber, Transcr., I, 355.) But it is of an earlier date, as we find two notices of it in the Stationers' Registers 1562/3, the first of which is: 'Recevyd of 'Rycharde applay for his lycense for pryntinge of a ballett intituled '*the complaynte of a synner vexed with payne*'. (Arber, *ut sup.*, 205). Now, this is a moralization (still extant, but never republished in complete form) of Elderton's poem by W. Birch, and begins thus:

The God of love,
that sits above,
Doth know us, doth know us,
How sinful that we be.

These lines, it will be seen, come very close to the portion of the song repeated by Benedick.

The second notice of Elderton's poem will be found on the same page of the Stat. Reg., where 'the answere to the IIIIth ballett made 'to *the godes of love*' is booked.

Perhaps it was Elderton's song which was entered on the Stat. Reg. in 1564/5 (Arber I, 271) under the title: "The complaynte of a lover beynge vexed with payne &c.".

Moreover, we possess three other references or allusions of Shakespeare's time to our ballad. In 'A Handefull of pleasant delites',

[1] Bullen, 'Old Plays', vol. I, p. 125.

[2] or of some later version or adaptation?—I only throw out this as a mere *possibility*.

[3] see Hazlitt, Coll. and Notes (s. v. Birch) p. 38.

1584, we find (p. 36 ed. Arber): 'The ioy of Virginitie: to, *The Gods 'of love'*, the first lines of which are precisely in the same measure as the verses in Much Ado referred to above,—

I Iudge and finde,
how God doth minde,
to furnish, to furnish
his heavenly throne above.

In 'Bacchus' Bountie' &c, *1593*,[1] is a song, beginning—

'The gods of love,
Which raigne above.'

The third allusion occurs in Th. Heywood's 'Fayre Mayde of the Exchange' (pr. 1607):

Ye gods of Love that sit above, and pitty Lovers paine.
Looke from your thrones upon the mones, that I do now sustaine.

GREEN SLEEVES.

The earliest mention of 'Green Sleeves' is of the year 1580. This dance tune to which a number of ballads were sung is printed in Chappell, I, 239 and by Naylor, 194. 'A new Courtly Sonet, of the 'Lady Greensleeves. To the new tune of Greensleeves', brought out in 1584 in 'A Handefull of pleasant delites' (Arber, Engl. Sch. Libr., III, p. 17), apparently contains something of the original song. Shakespeare, in The Merry Wives, twice makes mention of the tune. In Act II, I, 62, Mrs. Ford says that Falstaff's disposition and the truth of his words "do no more adhere and keep place together, than "the Hundredth Psalm to *the tune of 'Green Sleeves'*". Compare Mr. Naylor's comment on this passage (p. 75). In Act V, v, 20, Falstaff is made to say:

Let the sky rain potatoes; let it thunder to *the tune of Green Sleeves* . . . . . I will shelter me here.

'HAVE I CAUGHT MY HEAVENLY JEWEL'?

This 'Second Song', beginning thus, in Sidney's Astrophel and Stella (first printed 1591) is alluded to by Shakespeare in The Merry Wives (III, III, 45), where he makes Falstaff say to Mrs. Ford,

Have I caught thee, my heavenly jewel?[2]

[1] Reprinted in the Harleian Miscellany, vol. II, 1809, 8⁰, see p. 274.

[2] In the first and second Quartos the line stands without the *thee* of the Folio.

HEART'S EASE.

We find Heart's Ease mentioned as early as 1560, in Misogonus, as the tune of a song. See Chappell, I, 97. The original words are not known. It is referred to in 'Romeo', IV, v, 102:

> *Peter.* Musicians, O, musicians, *'Heart's ease, Heart's ease'*: O, an you will have me live, play *'Heart's ease'*.
>
> *First Mus.* Why *'Hearts ease'?*
>
> *Pet.* O, musicians, because my heart itself plays *'My heart is full of woe'*: O, play me some merry dump, to comfort me.[1]

HEIGH HO! FOR A HUSBAND.

In Much Ado, II, I, 332, Beatrice says:

> I may sit in a corner and cry *heigh-ho for a husband.*

'This, as Malone points out, is the title of a song in the Pepysian 'Collection at Magdalene College, Cambridge (vol. IV, p. 8): Hey ho, 'for a Husband. Or, the willing Maids wants made known. It is 'referred to again in III, IV, (ll. 54—55), and in Burton's Anatomy of 'Melancholy (ed. 1651, p. 565), part 3, sec. 2, mem. 6, subs. 3: "Hai-'"ho for an husband, cries she, a bad husband, nay the worst that '"ever was, is better than none".' (Wright, Clar. Press ed. of Much Ado, p. 105.) A song 'Oh! for a Husband' is reprinted, in part, with the music by Chappell, orig. ed. p. 454 from a MS. dated 1659. The burden of some stanzas of this song (which is perhaps a later version of the original) is:

> Oh! Oh! Oh for a husband!
> Oh! Oh! Oh for a husband!
> Still this was her song,
> I will have a husband, have a husband,
> Be he old or young.—

'*Oh! for a husband* is included in "A Complete Collection of Old and 'New English and Scotch Songs, with their respective tunes pre-'fixed", I, 91, 1735, and in all the editions of "Pills to purge Me-'lancholy", but there reset by Akeroyde'. Chappell, orig. ed., 782.

THE HUNT IS UP.

'The hunt is up' or simply 'hunt's-up' is the name of an old song and its tune, sung or played to awaken huntsmen in the morn-

[1] On 'My heart is full of woe' see *post*, p. 180, 'A pleasant new ballad of Two Lovers'.

ing, and also used as a dance. 'It seems to have been almost, if 'not quite, the most popular of the old ballad tunes'. The expression 'hunt is up' came to be used allusively to mean a song sung or tune played to rouse any one; an early morning song (see Chappell, I, 86, New Engl. Dict., s. v. hunt's-up). In this sense it is employed by Shakespeare in 'Romeo', III, v, 34:

Hunting thee hence with *hunt's-up* to the day.

Compare, too, Tit. Andron., II, II, 1 seq.—

The hunt is up, the morn is bright and grey, etc.

I CANNOT COME EVERY DAY TO WOO.

Of this song, 'very popular in the early part of Henry VIII.'s reign', the first stanza is to be found in 'an ancient MS.' from which John Stafford Smith reprinted both words and music in his 'Musica Antiqua', 1812. The song passed through a new phase in 1591—2, when it was entered on the Stat. Reg. as "a newe Ballad of *John wooinge* "*of Jo[a]ne* &c." (Arber Transcr., II, 602.) It is from this, in all probability (says Chappell), that the ballad in the Roxburghe Collection (II, 74) called 'The Country-man's Delight; or, The Happy Wooing. 'Being the successful Love of John the Serving-man, in his courting 'of Joan the Dairy-maid' has been copied. The burden 'I cannot come 'every day to woo' may have enjoyed proverbial currency. It is alluded to by Puttenham in his Arte of English Poesie, 1589, (Arber's Repr., p. 213) and, what is to our purpose, by Petruchio in The Taming of the Shrew, II, I, 115—'6:

Signior Baptista, my business asketh haste,
And every day I cannot come to woo." etc.

For more information about the song see Ballad Society, No. 19, Roxb. Ball., vol. III, p. 590-'6 [Compare, too, Bibl. Lindes., No. 576, and Rimbault, Songs and Ballads, 1851, p. 60.]

LIGHT O' LOVE.

This old tune is mentioned twice by Shakespeare. In The Two Gentlemen of Verona, I, II, 83, we read:

Julia: Some love of yours hath writ to you in rhyme.
Lucetta: That I might sing it, madam, to a tune.
Give me a note: your ladyship can set.
Jul. As little by such toys as may be possible·
Best sing it *to the tune of 'Light o' love'*.
Luc. It is too heavy for *so light a tune.*

Again, in Much Ado, III, IV, 41f.:

> *Hero.* Why, how now? do you speak in the sick tune? [*vide post*, p. 180, 'Sick, sick'.]
> *Beat.* I am out of all other tune, methinks.
> *Marg.* Clap's into '*Light o' love*'; that goes without a burden: do you *sing* it, and I'll *dance* it.
> *Beat.* Ye light o' love, with your heels! etc.

From this passage 'it appears that *Light o' Love* was strictly a *ballet*, 'to be sung and danced' (Chappell, old. ed., 222). The music will be found in Chappell, I, 82. The words of the original song are still undiscovered.

Regarding 'MAD TOM' cf. *ante*, s. v. Ovid (pp. 24-25).

MONSIEUR MINGO.

In 2. Henry IV. (Act V, III, 77—79) Silence sings:

> Do me right,
> And dub me knight:
> Samingo.

This is from an old song, beginning: 'Monsieur Mingo for quaffing 'doth surpass', to be found in Nash's 'Summer's Last Will and Testament' (1592; pr. 1600—see Dodsley, VIII, 55, 59, 61). Nash gives 'Domingo' instead of Silence's 'Samingo'.

MY MIND TO ME A KINGDOM IS,

a song in praise of contentment and humbleness, composed by 'E. Dier', seems to have been a favourite poem in the sixteenth century. It is in Percy's *Reliques* and in Arber's *Anthologies*. There is some slight reason to suppose that Shakespeare had it in view in 3. Henry VI., Act III, I, 59f.:—

> *Sec. Keeper*: . . . thou talk'st as if thou wert a *king*.
> *K. Henry:* Why, so I am, *in mind;* and that's enough.
> *Sec. Keep.* But, if thou be a king, where is thy crown?
> *K. Hen.* My crown is *in my heart*, not on my head;
> . . . . my crown is called *content.*

Sir Edward Dyer's poem[1] ought to be compared with this passage. Regarding the music, see Rimbault, *ut sup.*, p. 16.

[1] Set to music in 1588. (J. W. E.) The word 'slight' is J. W. E.'s.

MY ROBIN IS TO THE GREENWOOD GONE;
or,
BONNY SWEET ROBIN.

One tune to this song is at any rate older than 1597. 'The latter 'of the two versions given in William Ballet's Lute Book (*circa* 1600) 'is headed "Robin Hood is to the greenwood gone"; it is possible, 'therefore, that the original ballad was a song of Robin Hood.' This title must be the first line of the song. Another title is 'Bonny sweet 'Robin'. 'Nothing more is known of the words, unless the line sung by Ophelia in *Hamlet* (IV, v, 187),—

"For bonny sweet Robin is all my joy,"

'should be part of them, which, indeed, seems very probable.' (Chappell, I, 153.) "A ballad intituled A Dolefull adewe to the last Erle of "Darby. to the tune of 'Bonny sweete Robin'" was entered on the Stationers' Registers, 26 April, *1594*. (Arber, Transcr. II, 647.)

O DEATH, ROCK ME ASLEEP.

'Then death rock me asleep' is quoted by Pistol in 2. Henry IV. (Act II, IV, 211). The song, 'attributed with great improbability to 'Anne Boleyn, and perhaps with as little likelihood to her brother 'Viscount Rochford' (Dyce) is very old. 'The first stanza of the words 'with the tune, is contained in a Manuscript of the latter part of 'Henry VIII.'s reign', (Chappell, 2nd ed., p. 238.) Compare, too, Chappell, 3rd ed., I, p. 111, and Halliwell, ed. of Shakesp., vol. X, p. 112.

O MISTRESS MINE

is the name of a tune printed in Morley's *Consort Lessons*, 1599, which book, being for instruments, does not contain words. Feste's 'love-song in Twelfth Night, II, III, 40, exactly suits this air, reproduced by Chappell, I, 103, who remarks[1]: 'As it [the tune] is to be found 'in print in 1599, it proves either that Shakespeare's *Twelfth Night* 'was written in or before that year, or that, in accordance with the 'then prevailing custom, *O mistress mine* was an old song, introduced 'into the play.' 'The latter supposition is doubtless the true one,' says Dyce. Mr. Furness expresses his flat dissent from such a supposition, which *may* none the less be correct. A third alternative, however, still remains: Shakespeare may have remodelled an old song, or composed entirely new words for an older tune.

[1] Chappell, 1855-9, I, 209.

O SWEET OLIVER.

This song, now lost, is quoted by Touchstone in As You Like It, III, III, 101 seq:—

not,—

O sweet Oliver,
O brave Oliver,
Leave me not behind thee:

but,—

Wind away,
Begone, I say,
I will not to wedding with thee.

'It would seem that Touchstone is citing two distinct portions of the 'ballad' (Dyce)—or of two different ballads.—In the Stationers' Registers, Aug. 6. 1584 (Arber, II, 434), we find the following entry:

Ric. Jones.—Receaved of him for his license to printe A Ballat of. *O swete Olyver Leave me not behind the*[*e*] . . . . IIII d.

(Comp. Furness, VIII, 190, and Malone, VI, 449—451, for more information. See also Chappell, I, 88-9.)

THE PASSIONATE SHEPHERD TO HIS LOVE,
beginning,
COME LIVE WITH ME, AND BE MY LOVE,

'that smooth song which was made by Kit Marlow', is quoted, or rather misquoted, by Sir Hugh Evans in The Merry Wives (III, I, 17—29). Compare Chappell, I, 123. The poem was first published, fragmentarily, in The Passionate Pilgrim in the year 1599, and in complete form in England's Helicon, 1600.

PEG-A RAMSEY.

Malvolio's a Peg-a-Ramsey,

says Sir Toby in Twelfth Night, II, III, 81, referring to a well-known dance tune still preserved, and reprinted by Chappell (I, 248). It, together with Green Sleeves and other dance tunes, is mentioned by Nash in 'Have with you', etc., 1596 (Huth Libr. ed., III, 181). No earlier version of a 'Peggy Ramsey' ballad is known than the song in 'Wit and Mirth or Pills to purge Melancholy' (1707, 1714, vol. III, 219; or 1719, V, 139). Being amazingly indecorous, it has seldom been reprinted. [Burns made use of this or some other version for his 'Cauld is the e'enin blast'. See Cent. Ed., III, 203, 444.]

A PLEASANT NEW BALLAD OF TWO LOVERS,

TO A PLEASANT NEW TUNE,

is the title of a song, beginning 'Complaine, my lute, complaine on 'him,' which has been handed down to us in a copy 'of the commence- 'ment of the seventeenth century.' The first stanza ends with the following line: 'Hey, hoe! *my heart is full of woe!*' It is thought that this song (reprinted by the old Shakesp. Soc.,—Papers, 1844, p. 13, and by the Ball. Soc. No. 9, Roxb. Bds., vol. II, p. 305) is the very one Shakespeare quotes, in 'Romeo', IV, v, 107 (*vide ante* 'Heart's Ease', p. 175).

SICK, SICK.

This is the name of two old tunes printed by Chappell (I, 73-74), who remarks: 'In Much Ado about Nothing, [Act, III, iv, 41] Hero 'says, "Why, how now? do you speak in the *sick tune*?" and Beatrice 'answers, "I am out of all other tune, methinks."' An old 'ballad 'in a handwriting of about the last quarter of the sixteenth century, 'which may perhaps be the original to which these tunes belong' is printed, in part, by Chappell *(ut sup.)*.

A SONG TO THE LUTE IN MUSICKE;[1]

or,

IN COMMENDATION OF MUSICKE,

beginning: 'Where gripinge grefes the hart would wounde.' 'This 'song, both words and music, is the production of Richard Edwardes' (Rimbault). The words were printed in 'The Paradise of Dainty Devises', which first appeared in 1576 and passed through 8 editions within 24 years. The first lines are quoted (but with several variations), and humorously commented on, by Peter in 'Romeo', IV, v, 128 *seq.* Reprints of the song are numerous. Percy and Rimbault give the song from MS. copies.

THERE WAS AN OLD FELLOW AT WALTHAM CROSS.

Speaking of the wildness of his youth Shallow says: "our watch- "word was 'Hem boys!'" (2. Henry IV., Act III, ii, 231). Staunton observed very pertinently that "With a hem, boys, hem, And a cup

[1] The first title is that given in a 'MS. volume of old tunes, etc. which purports to be "Thomas Mulliner's Boke for ye Organ or Virginalls."' Rimbault, Mus. Ill. of Percy's Rel., p. 7. The music will be found there on p. 49. The song is not given by Chappell.—'In Commendation', etc. is the heading in 'The Paradise'.

of old sack" is, apparently, the refrain of the above song.[1] "This is "quoted as an old song in Brome's play, *The Jovial Crew* .... acted 1641. "It is also in the *Antidote against Melancholy*, 1661." (Chappell, II, 158, where also the air is given.)

WHERE IS THE LIFE THAT LATE I LED?

This song referred to by Petruchio (Shrew, IV, I, 143)[2] and by Pistol (2. Henry IV., Act V, III, 146) is no longer extant. But we know in what metre the poem was written, and what subject it dealt with, and when it was probably composed. First, there is a song *to the tune* of 'Where is the Life that late I led' in A Gorgeous Gallery of Gallant Inventions (*1578*). This settles the metre question. Then we have a poem in a Handefull of pleasant delites (*1584*) entitled 'Dame Beauties *replie* to the Lover late at libertie: and now complaineth 'himselfe to be her captive, Intituled: Where is the life that late I 'led.' (Arber, Engl. Sch. Libr., III, 14.) This *reply* to the lost song allows us to form some idea as to the contents of the latter. The title just given and the general contents of the '*reply*' harmonize with the title of what may be the original song: 'a *newe* ballet of one 'who myslykeng his lybertie soughte his owne bondage through his 'owne folly'. This is an entry on the Stationers' Registers dated 1565/6 (Arber, Transcr., I, 308.)

WHOOP, DO ME NO HARM, GOOD MAN.

This song is alluded to by Shakespeare in The Winters Tale, IV, IV, 199—201. In 'The Famous History of Friar Bacon' there is a ballad to the tune of "*Oh* doe me no harme good man." (Thoms, Early Prose Rom., 1858, vol. I, p. 224.) 'A song [to this tune "Whoop, '"do mee", etc.] will be found in Fry's Ancient Poetry, but it would 'not be desirable for republication,' says Chappell (orig. ed., 208). *Pourquoi pas?* Its date is probably *c*. 1615. Ford, in Act III, sc. III, of "The Fancies chaste and noble" (pr. 1638) places the line "*Whoop,* "*do me no harm, good woman*" in the mouth of Secco.—The music (1610) is in Chappell (I, 96).

WILLOW, WILLOW,

is Desdemona's swan song (Oth. IV, III). Shakespeare, in making use

[1] In An Antidote against Melancholy (1661) it is called a catch (=round).

[2] "Where is the life that late I led? Where are those—Sit down, Kate, etc." I am inclined to regard "Where are those" [pleasant days?] as the continuation of the song. ("No doubt", says J. W. E.)

of the words of the old song ("an old thing 'twas", l. 29) 'has made 'changes which were necessary to suit them to a female character.' A later version of the song is in the Roxburghe Collection, Ballad Soc., 4, Roxb. Bds. I, p. 171, a version, which is nearly identical with that printed by Percy in his Reliques from the Pepys collection. Each is of the first half of the seventeenth century,[1] but the former is rather to be preferred. But an *earlier* copy (with the music) than either is to be found reprinted in Chappell, Old Engl. Pop. Music, I, 106; but it does not everywhere show the greater agreement with Shakespeare's version. (Cp., too, Chappell, original ed., 206, 774.)

ROUNDS.

JACK, BOY! HO! BOY!

This round for four voices, is printed in Th. Ravenscroft's Pammelia, 1609, the first collection of popular rounds printed in England. The words are as follows:

Jacke, boy, ho, boy, newes!
The Cat is in the well.
Let us ring now for her knell,
Ding, dong, ding, dong, bell.

(see Knight, Pict. Shaksp., Comedies I, p. 316, where the music is given). The first line is alluded to in '*The Shrew*', IV, I, 41:—

Curtis: "There's fire ready; and therefore, good Grumio, *the news.*
Gru. Wy, '*Jack boy! ho! boy!*' and as much *news* as will thaw.
Curt. Come, you are so full of cony-catching!

THOU KNAVE.

In Twelfth Night, II, III, Sir Toby having made the proposal to sing a catch, that is, a round or roundelay, Sir Andrew, says (v. 66): "Let our catch be, 'Thou knave'".—*Clown*: "'Hold thy peace, thou "knave', knight? I shall be constrained in't to call thee knave, knight." *Sir Andrew:* "'Tis not the first time I have constrained one to call "me knave. Begin, fool: it begins 'Hold thy peace'." This catch 'Hold thy peace' was printed in the 'Deuteromelia', a work supple-

[1] This is Chappell's opinion, (Ball. Soc., *ut sup.*). Mr. Wooldridge, however, who prepared the latest edition of 'The Pop. Mus.' thinks (I, 109) the version, as found in the Roxburghe Collection, falls after the Restoration. But Mr. Wooldridge must be in the wrong; as the Roxburghe copy is printed for Edward Wright, whose dates are 1620—1655; see Roxb. Ball. I, Ball. Soc., p. 174 and p. XXIII.

mentary to the 'Pammelia', both published by Th. Ravenscroft in 1609, when he was about 17 years of age. The music, reproduced by Hawkins (see Furness, XIII, 118), is beyond doubt original.

THREE MERRY MEN.

The original words are probably the following, given by Peele in his 'Old Wives' Tale,' 1595:

let us rehearse the *old* proverb [=song]—

'Three merry men, and three merry men, and three merry man be we:
'I in the wood, and thou on the ground, and Jack sleeps in the tree.'

The melody as it is preserved to us by Playford, *circa* 1650, is given by Chappell, I, 197, and, Naylor, 189. I have no doubt but that 'Three Merry Men' was originally a round, or two rounds, for three voices. The three Parts would suit the words as follows:

1st	voice	Three merry men and
2nd	„	Three merry men and
3rd	„	Three merry men be we.
1st	„	I in the wood and
2nd	„	Thou on the ground and
3rd	„	Jack sleeps in the tree.

The song is alluded to by Sir Toby in Twelfth Night, II, III, 81: 'Three merry men be we'.

POPULAR RHYMES.

1) In The Merry Wives (IV, II,) we hear that, Mr. Ford, considering himself a cuckold, "buffets himself on the forehead crying, "'Peer out, peer out!'" (l. 25),—that is, appear horns! Henley thinks that 'Shakspeare here refers to the practice of children, when they 'call on a snail to push forth his horns:

'Peer out, peer out, peer out of your hole,
'Or else I'll beat you black as a coal.'

---

2) *Pillycock, Pillycock, sat on a hill;*
If he's not gone, he sits there still.

These lines are given by Ritson in his collection of mursery rhymes entitled 'Gammer Gurton's Garland, or the Nursery Parnassus', 1783. Compare King Lear, III, IV, 78:

Edgar: "Pillicock sat on Pillicock-hill".

---

3) *When Adam delved and Eve span,*
Who was then a gentleman?,

is alluded to by the clown in Hamlet, V, I, 32f.:—

First Clown: .... Come, my spade. There is no ancient gentlemen but gardeners, ditchers, and grave-makers: they hold up Adam's profession.
Sec. Clo. Was he a gentleman?

Regarding this old rhyme compare Hazlitt 'English Proverbs', 1869, p. 455. We find it quoted by Greene: 'I will not forget the old wives 'logick, when Adam delvd and Eve spanne, who was then a Gentleman?' (A Quip for an Upstart Courtier, 1592. Huth Libr., XI., p. 225.)

The couplet is also to be found in Holinshed (cf. Bosw.-Stone, p. 272, n. 2), who relates that John Ball, the fomenter of Wat Tyler's Insurrection (1381), made it the theme of his sermon at Blackheath. There is an evident allusion to this in 2. Henry VI., Act IV, II, 142:

Sir Humphrey Stafford: Villain, thy father was a plasterer;
And thou thyself a shearman, art thou not?
Cade. And Adam was a gardener.

Compare also John Holland's assertion, ll. 9—10:

Well, I say it was never merry world in England since gentlemen came up.

---

4) "For I the ballad[1] will repeat,
Which men full true shall find;
Your marriage comes by destiny,
Your cuckoo sings by kind."
(All's Well, I, III, 64.)

Something like two of these lines are to be found in John Grange's 'Garden', *1577*[2]:

'Content your selfe as well as I, let reason rule your minde,
'As Cuckoldes come by destinie, so Cuckowes sing by kinde.'

---

[1] ballad='a proverbial saying, usually in form of a couplet'. (New Engl. Dict.)
[2] On folio R.ij.

A Spell.

In King Lear (III, IV, 125ff.) we have the following spell:—

S. Withold footed thrice the old;
He met the *night-mare*, and her nine-fold;
Bid her alight,
And her troth plight,
And, aroint thee, witch, aroint thee!

In illustration of this truly Teutonic spell, with the epic introduction and the winding up with the charm proper, we might cite numerous examples, beginning with the 'Merseburger Zaubersprüche', the oldest of them. A characteristic charm against the nightmare or Incubus, coming near to the one quoted by Edgar, is to be found in Scott's 'Discoverie of Witchcraft', 1584, (reprinted, 1886,—see p. 68 *ib.*) and deserves to be quoted here:

S. George, S. George, our ladies knight,
He walkt by daie, so did he by night:
Untill such time as he hir found,
He hir beat and he hir bound,
Untill hir troth she to him plight,
She would not come to hir that night.

FURTHER NOTES AND COMMENTS ILLUSTRATIVE OF SOME OLD SONGS AND BALLADS IN SHAKESPEARE'S PLAYS.

The song in Twelfth Night, II, IV, 52,

"COME AWAY, COME AWAY, DEATH", etc.

appears to be a later interpolation and not the original song intended by the great dramatist, for it cannot be said to chime in with the description of it given immediately before[1]:

"it is old and plain;
"The spinsters and the knitters in the sun
"And the free maids that weave their thread with bones
"Do use to chant it: it is silly sooth,
"And dallies with the innocence of love,
"Like the old age."

[1] J. W. E. objects strongly (see below, Addenda and Corrigenda).

Dolphin my Boy.

The words "*Dolphin my boy, my boy, sessa! let him trot by*", (Lear, III, iv, 104) are declared by Steevens to be part of a ballad. But his note sounds too romantic (see Furness, V, p. 192). "He shall be Dauphin my boy" is a passage in Ben Jonson's Bartholomew Fair (V, iii).

HOW SHOULD I YOUR TRUE LOVE KNOW?

The three stanzas quoted by Ophelia in Hamlet, IV, v, 23f., show a marked affinity with (being also in the same measure as) a ballad beginning 'As you came from the holy-land', reprinted by Percy in his Reliques. A better copy is in The Garland of Good-Will (Percy Soc., XXX, p. 111) and in Percy's MS. (III, 471).[1] The substance of this ballad is given in the following words by Mr. Hales (III, 465): 'A lover growing or grown old, it would seem, has been left in the 'lurch by the object of his affections. As all the world thronged to 'Walsingham, the lover supposes she too must have gone that way; 'and meeting a pilgrim returning from that English Holy Land, asks 'him if he has seen anything of her runaway ladyship. The lover, 'having described how his true and untrue love may be known from 'many another one, learns that she has been met making for Wal-'singham; and then, asked why she has deserted him, explains that, 'though she once loved him, she has lost her love now he waxes old, 'and generally, that a woman's love is ever capricious and veering' &c. The situation is somewhat similar in Ophelia's fragment, where we must suppose the lady to enquire of a pilgrim whether he has met with her lover, to which he replies "How should I your true love "know From another one?" (Compare: 'How should[2] I know your 'true love, That have met many a one?'—first half of stanza 2 of the above mentioned ballad). She tells him by what tokens, and learns that he is dead. What the cause of his death is we do not ascertain from the fragment. But we cannot be far wrong in supposing that some clue is to be found in 'Gentle Herdsman' (in Percy), where a lover is killed by the scorn of a lady, who now repents when it is too late and pilgrims to Walsingham.—After having taken every circumstance into consideration, I am inclined to hazard the conjecture that Ophelia's fragments (ll. 23—39) belong to the original 'Walsingham' song, now lost.

[1] Arber has also reprinted it in his 'Anthologies' from MS. Rawl. Poet.

[2] Rawl. MS., *shall*.

JOG ON, JOG ON.

The tune with this title is preserved in the 'Dancing Master', 1650, etc., and with the title Hanskin in the Fitzwilliam Virginal Book (early 17th cent.) and with a third in Pills to purge Melancholy 1707, etc. Besides the stanza sung by Autolycus, Wint. Tale, IV, III, 132, two more a given in the Antidote against Melancholy, 1661, no earlier copy of them being known. (see Chappell I, 159, for the words and tune).

THE MAN SHALL HAVE HIS MARE AGAIN.

The man shall have his mare again, and all shall be well.

(Mids. N. D., III, II, 463.) See a note in Ballad Soc. No. 37 p. 746. Referring to Roxburghe Ballads, vol. I, p. 57, l. 29 where 'The man 'shall have his Mare agen' occurs, the Rev. Mr. Ebsworth observes: 'Compare *Midsummer Night's Dream*, Act [III], where the same line 'is quoted from an older ditty of *How Mosse found his Mare.* She 'was caught napping, and it is proverbial.' I always regarded the phrase as being proverbial. I can give no information about 'the 'older ditty'. Dr. Ritter directed my attention to the above passage in the Ball. Soc. Compare also Wright's note (Clar. Press, or in Furness).

"O THE TWELFTH DAY OF DECEMBER",

is a line put into the mouth of Sir Toby (Twelfth Night, II, III, 90). Walker conjectures O' for O, which sounds plausible enough. Kittredge suggests that the ballad quoted by Sir Toby may be 'Musselburgh Field' (Child, III, 378; IV, 507), which celebrates the Battle of Pinkie fought in 1547, and begins thus:

On the *tenth* day of December.

(Stanza 5 begins:

On the *twelfth* day in the morne)

This is not *certain.* There is a ballad of the year 1584, reprinted in Ancient Ballads and Broadsides (Lilly, 1870, p. 182) beginning,

The twelfe day of November last.

"SLEEPEST OR WAKEST THOU, JOLLY SHEPHERD?"

(Lear III, VI, 43–46). Regarding the question 'sleepest or wakest?' compare Child, II, 240.

'TAKE, O, TAKE THOSE LIPS AWAY', etc.

(Meas. f. Meas. IV, I). The song recurs in the 'Bloody Brother' (Fleay:

c. 1616; Ward: after 1624), by Fletcher, etc., with an additional stanza, which is evidently by another and inferior hand. The second stanza sounds like a burlesque of the first, which may after all be Shakespeare's own composition.

TO-MORROW IS ST. VALENTINE'S DAY.

Of this apparently old song consisting of four stanzas *sung* by Ophelia (Hamlet, IV, v, 48-55, 58-65) nothing further is known. Douce found a parallel to the lines

> Let in the maid, that out a maid
> Never departed more—

in a French ballad of 1598:

> 'Elle y entra pucelle
> 'Grosette elle en sorta.'[1]

WAS THIS FAIR FACE, etc.

> "*Was this fair face the cause, quoth she,*
> "*Why the Grecians sacked Troy?*" *etc.*
> (All's Well, I, III, 74-83.)

A *ballad*, called 'The lamentations of Hecuba and ye ladies of Troye', was entered on the Stat. Reg. in 1586 (Arber, Transcr., II, 451).

WE WILL BE MARRIED O' SUNDAY.

In the Taming of the Shrew (II, I) Petruchio, having decided to wed Kate upon Sunday (l. 300), takes his leave saying (l. 324):

> "Sunday comes apace:
> "We will have rings and things and fine array;
> "And kiss me, Kate, *we will be married o' Sunday.*"

In illustration of this passage, Halliwell gives the following exhaustive note, which I quote verbatim:

We will be married o' Sunday. Petruchio is here probably quoting from some old ballad. The earliest song with a similar burden is one in Ralph Roister Doister, 1566, which commences,

---

[1] The reader of Goethe's Faust will be reminded of

> *Mephisto* . . . 'Lass, lass es sein!
> 'Er lässt dich ein,
> 'Als Mädchen ein
> 'Als Mädchen nicht zurücke.'

in the scene Strasse vor Gretchens Thür.

I mun be maried a Sunday;
I mun be maried a Sunday;
Whosoever shall come that way,
I mun be maried a Sunday.[1]

There is a ballad of the last century, which may be a modernized version of an earlier one, commencing,—

As I walk'd forth one May morning,
I heard a fair maid sweetly sing,
As she sat under her cow milking,
We will be married o' Sunday.[2]

Another stanza of which may be quoted, as illustrative of the belief that Petruchio's speech refers to a ballad,—

Then on my finger I'll have a ring,
Not one of rush, but a golden thing;
And I shall be glad as a bird in spring,
Because I am married o' Sunday.

The present ballad is either copied, or is connected in some way, with "a country song" which is introduced into Mrs. Centlivre's comedy of the Platonic Lady, 1707, which commences as follows,—

As I walk'd forth one May morning,
I heard a pretty maid sweetly sing
As she sat under the cow a milking,
Sing I shall be marry'd a Tuesday;
I mun look smug upon Tuesday.
(Halliwell's Folio ed. of Shakesp., 1856, vol. VI., p. 391.)

Collier, in his edition of Shakespeare, 1842, vol. III, p. 148, gives verses[3] from oral tradition, which come marvellously near Shakespeare's words. But I suspect, Collier is swindling.

BURDENS.

In Hamlet, IV, v, 170, Ophelia says, "You must sing '*a-down* "*a-down*', And you '*call him a-down-a*'." This, I take it, is the way to print this passage. Ophelia desires the by-standers to sing the burden. Again, in The Merry Wives (I, IV, 44) Mrs. Quickly sings: "*And down, down, adown-a, &c.*" The words 'down, down', etc., as

[1] See Dodsley's old Engl. Plays, 1874, vol. III, 159.
[2] The whole ballad is reprinted in Old Sh. Soc., Papers, 1844, pp. 80—82.
[3] also reproduced by Delius.

a refrain or undersong, are to be met with rather frequently in old songs and ballads.[1]

In The Winter's Tale (IV, iv, 195) Autolycus is said to have songs "with such delicate burthens of *dildos* and *fadings*, 'jump her "and thump her'", etc.—*Dildo* occurs as a refrain in old songs. It also had a coarse meaning (='mentula factitia. *Cotgrave* in *Godemiche.*' Wright, Prov. Dict., Cp. New Engl. Dict.). The curious reader will find full information concerning this burden in Ballad Society, No. 16 (Bagford Ballads, vol. I) p. 551,—J. W. Ebsworth's introductory remarks to a coarse poem: 'The Maid's Complaint for Want of a Dil doul'.

Fading was the name of a dance and 'With a fading' the name of a tune, several songs to which with this burden are known (see Chappell, II, 104; old. ed., 235). A Round of Matt. White, 'The Cour-'tier scorns the country clowns' (date about 1600) has for its third and last line 'With a fading, fading, fading, fading', etc. (Naylor, 82).

The burden *'hey nonny, nonny'*, etc. (Much Ado, II, iii, 71; Haml., IV, v, 165; As You L. It, V, iii, 18, etc.; Cp. Lear, III, iv, 103) is met with in older songs; e. g., in Chettle's Old Grissill is a song, the first, with such a burden; also in The Two Noble Kinsmen (III, iv) and elsewhere. Coverdale refers to this burden (see Chappell, old. ed., 53-4, or Clar. Press ed. of As You L. It, p. 160).[2]

A somewhat similar refrain occurs in As You Like It, V, iii, 21, etc., "*hey ding a ding, ding*". This is a burden in 'Old Simon the King' (Chappell, I, 280) and is quoted as the title of a song by Laneham (Cp. Furnivall's Introd.).

A NOTE.

To ballads on strange fishes and on monstrosities such as are

[1] Comp. Chappell, old. ed., pp. 59, 219, 222, 376, 391; and Child, III, Robin Hood ballads. Ophelia's burden 'will be found almost verbatim in a ballad 'commencing—

'When as King Edgar did govern this land,
'*Adown, adown, down, down, down,*
'And in the strength of his years he did stand,
'*Call him down-a.*'

Chappell, old. ed., p. 768. This ballad is reprinted by the Percy Soc, XXX, 12. Another song to the same tune is in 'A Handefull of pleasant delites', 1584, (Arber, Engl. Sch. Libr., III, 57, cp. also p. XIII). A third in 'Misogonus', *circa* 1560, (Brandl, *Quellen*, p. 456). A fourth in Anc. Ball. and Broads., Lilly, 1870, p. 78, and compare pp. 288-9. No doubt the burden formed an essential part of the tune, 'Labandalashot', for such is the name of it.

[2] Comp. Anglia, XII, 236.

spoken of by Autolycus in Wint. Tale, IV, iv, Halliwell devotes five folio pages and a full-page illustration, which the reader may turn to if he is curious to know something more about these 'stretchers'. Examples of ballads of this sort will be found in 'Ancient Ballads and Broadsides', Lilly, 1870.

POPULAR TALES AND LIGHT LITERATURE.

A HUNDRED MERRY TALES.

'This is one of the best of our old Jest-Books, and is alluded 'to by Shakspere in his Much Ado about Nothing',[1] II, i, 134-6:

> *Beat.* That I was disdainful, and that I had my good wit out of the 'Hundred Merry Tales':—well, this was Signior Benedick that said so.

From this passage we may gather that Shakespeare, when he wrote his play, considered this production as rather trite. There are two editions of it extant, both by Rastell, and of each only one copy is known. The more perfect copy (1526) is reprinted by Oesterley (London, 1866); the other by Hazlitt. The book is entered twice on the Stat. Registers 1557/8 (Arber, I, 75) and 1582 (Arber, II, 405). The latter entry implies yet another edition.

THE JESTS OF SCOGAN.

In 2. Henry IV. (Act III, ii, 32) Shakespeare relates how Falstaff broke "Skogan's head at the court-gate, when a' was a crack not "thus high." There can scarcely be any doubt, that Shakespeare had in view Scogan, the buffoon, the hero (fictitious or otherwise) of a jest-book named after him, which was booked on the Stat. Reg. in 1565/6 (Arber, I, 299). But the earliest edition now known is dated 1626. For more information see Dict. of Nat. Biogr. The Jest-Book is reprinted by Hazlitt, *ut sup.*, vol. II.

ROBIN GOODFELLOW.

We possess abundant proof that tales of Robin Goodfellow's pranks (also called Puck or Hobgoblin) passed current among the people long before the date of Midsummer Night's Dream. One "E. D.", in his *Briefe and necessary Instruction*, 1572, mentions the 'tales of Robin Goodfellow', which, as the writer seems to imply, already existed in

[1] Furnivall, Introd. to Laneham's Letter, p. cviii.

print.[1] Whether these 'tales' were in prose or in verse we do not know. Perhaps both. What their contents must have been, may be gathered from the chap-books and ballads concerning Robin Goodfellow printed at a later date, collected and republished by Halliwell in old Shakesp. Soc., 1845, pp. 120f.

GILLIAN OF BRAINFORD'S TESTAMENT,

a humorous but coarse poem by Robert Copland, enjoyed a great deal of popularity in Elizabethan days. It was first printed shortly after 1562. Rob. Laneham mentions it in his list of popular books (see Furnivall, *ut sup.*). The poem was reprinted by Furnivall in 1871 (for private circulation, 'in order to avoid possible annoyance from any cantankerous puritan'). Nash, in his Preface to Greene's Menaphon (1589—Huth Libr. ed., Greene VI, 13) speaks of 'the idle usage of our unexperienst punies' into whose 'libraries' 'a tale of Ihon [= Joan?] a Brainfords will, 'and the unluckie furmentie[2], wilbe as soon interteined . . . ., as the 'best poeme that ever Tasso eternisht: which being the effect of an 'undescerning iudgement, makes drosse as valuable as gold.' Nash again alludes to her in Summer's Last Will (Huth Libr., Nash VI, 89). Another important allusion is in Westward Ho (*c.* 1605): 'I doubt 'that old hag, Gillian of Brainford, has bewitched me' (Webster, ed. Dyce, 1877, p. 238). A drama called 'fryer Fox and gyllen of Bran-'forde' is mentioned in Henslowe's Diary, 1598-9. And 'a pleasant 'newe Ballad of Julyan of Brainfords Last will and testament' is booked on the Stat. Reg. 1 March, 1600.

"The fat woman of Brentford" (Quarto: *Gillian of Brainford*"), "a witch" and "hag", is introduced in The Merry Wives. In Copland's production, however, no mention is made of her being addicted to witchcraft or of her fatness. Her legacy consisted of f—s left to some foolish folks.

THE BOOK OF RIDDLES.

Slender: "You have not *the Book of Riddles* about you, have you?"
Simple: "Book of Riddles! why, did you not lend it to Alice Shortcake upon All-hallowmas last, etc.?

(Merry Wives, I, I, 208)

This was 'The Booke of mery Riddles. Together with proper 'Questions, and Wittie Proverbs to make pleasant Pastime. No lesse 'usefull then behoovefull for any yong man or child to know if he be

[1] See Shakespeare Jahrbuch, 1904 (next year).

[2] furmenty=frumenty; accounted unlucky, because it caused the farts.

'*quicke-witted* or no.' London . . . . 1600 ($8^{vo.}$ 24 leaves).—'We can 'very well believe that this . . . . was the edition which Shakespeare 'had in his mind when he wrote "The Merry Wives", about the 'date when the *reprint* before us (for such it no doubt was) was brought 'out. We take it also, that it was a recent edition of the same "book 'of riddels" which Langham in his Letter from Kenilworth mentions 'in 1575'. (Collier, Bibl. Account of the Early Engl. Lit., vol. II, 264). No edition earlier than 1600 is known to us.

CHILD ROWLAND.

Those familiar with the story of Jack the Giant-Killer will recollect the words which are put into the mouth of a giant whose castle Jack enters:

Fe, fa, fum!
I smell the blood of an Englishman.

Some such formula appears to belong to the common stock of ogre and giant stories of the Indo Europeans (see Clouston, Popular Tales and Fictions ,1887, vol. I, 134). The above lines are quoted by Nash in 'Have with you', etc., 1596,[1] and alluded to by Peele in his 'Old Wives' Tale', 1595. And in King Lear (III, IV, 187) Edgar is made to say:

Child Rowland to the dark tower came,
His word was still—Fie, foh, and fum,
I smell the blood of a British man:

Now, the earliest known edition of the Jack and the Giants story is 1711.[2] In this comparatively late version of the story, which, I suppose, had passed through a good many metamorphoses, we find no explanation for "Child Rowland". Here a Scotch folk-story, printed by Robert Jamieson in the 'Illustrations of Northern Antiquities', 1814, p. 397 f., gives us some clue. Child Rowland, the hero of this tale (which is interspersed with poetry) having contrived to enter the castle of the King of Elfland is sniffed by the latter,

---

[1] Huth Libr., Nash III, 53.

[2] in the Brit. Mus.; but part 1. is wanting. On p. 16 we find the following verses:—

Fe, Fi, Fo, Fum,
I smell the Blood of an English Man;
Be he alive, or be he dead,
I'll grind his Bones to make me Bread.

Halliwell-Phillipps, referring probably to the same copy in the Brit. Mus., says the date 1711 must be a mistake for 1741 or 1771. (Folio ed. of Shakesp., vol. XIV, p. 465.)

With 'Fi, fi, fo, and fum!
'I smell the blood of a *Christian* man', etc.

This story evidently stands in close literary relationship to Jack the Giant-Killer on the one hand, and to three Danish ballads of 'Rosmer Hafmand' (translated by Jamieson) on the other.[1]

AN OLD TALE,

in which the words *'it is not so, nor it was not so, and God forbid it should be so'*, occurring also in Much Ado, I, I, 218, form a significant part of the story, was related by Blakeway and is to be found in many editions of the play. This, or some story related to it, may have been 'the old tale' referred to by Shakespeare (see Furness, XII, 29).

ABEL SLAIN WITH THE JAWBONE OF AN ASS;—AND MURDERED AT DAMASCUS.

According to an old legend, Cain slew his brother with the jawbone of an ass. Skeat[2] sees an allusion to this legend in Hamlet, V, I, 84—87:—

how the knave jowls it [the skull] to the ground, as if it were *Cain's jaw-bone, that did the first murder!* It might be the pate of a politician which this ass now o'er-reaches.

Cain was said to have committed the murder on the site where Damascus now stands. Compare 1. Henry VI., Act I, III, 39:—

This be Damascus, be thou cursed Cain,
To slay thy brother Abel, if thou wilt.

THE OWL WAS A BAKER'S DAUGHTER.

In Hamlet (IV, v, 41—44) Ophelia says:—

Well, God 'ild you! *They say* the owl was a baker's daughter. Lord, we know what we are, but know not what we may be. God be at your table!

'Douce was the first to point out that this probably referred to the 'legend that a baker's daughter, who refused to give bread to Christ, 'was transformed by the Saviour into an owl. But none of our anti-'quaries has, I believe, mentioned that in Cornwall the legend is 'familiar, and of old date' (Hazlitt, Pop. Antiq. of Gr. Britain, 1870, III, p. 196). Douce had observed that the story was common among the vulgar in Gloucestershire.

---

[1] I think there is an allusion to our Child Rowland story in Nash's Summer's Last Will *ad init.* (Huth Libr., Nash VI, 89).

[2] Notes and Queries, 6th S. II. Aug. 21, '80; and The Academy, Oct. 26, '95.

'According to Charles G. Leland (The English Gipsies, etc., 1873, 'p. 16) even the Gipsies are all familiar with the monkish legend.' see Elze's ed. of Hamlet, 1882, p. 213. Compare also Hazlitt, English Proverbs, 1869, p. 381.

SCHNEEWITTCHEN.

There are some striking resemblances between this famous German fairy-tale and Shakespeare's Cymbeline. While admitting these, I do not contend that the author of this play was familiar with the tale itself. All I dare assert is, that he was acquainted with some tale or tales distantly related with it.[1]

From

BEAST FABLES

are derived the following names of animals:

Chanticleer=cock. "Crow like Chanticleer" (As Y. L. It, II, VII, 30). "The strain of strutting Chanticleer" (Tempest, I, II, 385). *Partlett*=hen. In 1. Henry IV., Act III, III, 60, Falstaff addresses the Hostess as "Dame "Partlett the hen!" Compare also Wint. Tale, II, III, 75: "Thy dame "Partlett."—Chanticleer and Pertelot are familiar to the readers of Chaucer (Nun's Priest's Tale) and Caxton's 'Reynard the Fox'. *Tibert*= cat. Cf. 'Romeo', II, IV, 18: "*Benv.* Why, what is Tybalt? *Merc.* More "than prince of cats." Compare, too, III, I, ll. 78—81 & 104. Tibert occurs as the name of the cat in 'Reynard the Fox'. In Nash's 'Have with you', etc., 1596, we find: "Not *Tibault* or *Isegrim*; Prince of "Cattes" (Huth Libr., Nash, III, 74). In Dekker's Satiromastix there is another allusion, 'Tyber[t], the long-tail'd prince of Rattes' (*sic*).[2] Jonson speaks of cats as Tiberts (Epigr. 133).—As for Shakespeare's acquaintance with the *Æsopian* Fables, compare *ante*, pp. 17—19.

A NOTE.

"The humour of forty fancies" in the Shrew (III, II, 70) 'is ge'nerally understood to mean some collection of the short poems called '*Fancies*, which Petruchio had stuck into his lackey's hat' (Dyce). The supposition seems to me absurd. Fancy, pinning a book on to your hat! Why, it has no meaning. I rather think Shakespeare is making use of a bold figure of speech. Malone gives a quotation to show that 'fancy' could mean some ornament. The passage would therefore mean: the hat is fantastically and *humorously* trimmed with forty ornaments or ribbons.

[1] Compare Germania, IX, 458; Simrock 2nd ed., p. 274; Anglia, VI, 34f.; and Gollancz, Cymbel., in the Temple Shakespeare.

[2] Dekker's Dram. Work's, London, John Pearson, 1873, vol. I, p. 259.

CHAPTER 6.

THE BIBLE AND THE PRAYER BOOK.

ELIZABETHAN BIBLES.

Shakespeare's relations to the Bible have been dealt with at length though neither altogether exhaustively nor without diffusiveness by Bishop Charles Wordsworth, in his book "Shakspeare's Knowledge and Use of the Bible", to which I must refer the reader for information. Read also Dr. Furnivall's Forewords to "Shakspere and Holy Writ" (Marcus Ward).[1] From these works it will appear that Shakespeare's mind was thoroughly saturated with the Bible story, which seems as much part of him as his love of nature and music. What Shakespeare himself believed has been debated, but we can scarcely doubt that he accepted a good deal more than the 'advanced modern thinker'. The Deists, Hume, Kant, and Darwin had not yet appeared on the world's stage.

I do not desire here to go over old ground and repeat what has been said by others about Shakespeare's knowledge of Holy Writ. The present section has only to do with some questions which may be said to belong to higher criticism.

The two standard Bibles of Elizabeth's reign were the 'Geneva Bible' and the so called 'Bishops' Bible'. The former first published in complete form in 1560, became the household bible of the English people and no fewer than sixty editions of it passed into circulation during the reign of Queen Elizabeth. For many years it successfully competed even with the Authorised Version of 1611. The Bishops'

[1] Compare also J. B. Selkirk, Bible Truths with Shakspearian Parallels; and Colton, Shakspeare and the Bible.

Bible first printed in 1568 was the version authorised to be read in the churches and passed through nineteen editions. 'It may be 'doubted whether it was ever cordially received. The Great Bible of '1539 was still used in many churches, and the Geneva Bible was in 'almost every house.' (Encyclop. Brit., VIII, p. 388.) If this was the case, it is just possible that the Great Bible may have been used in a provincial parish church like that of Stratford. Moreover, we have to bear in mind that the Common Prayer-Book contained the Psalms and the Epistles and Gospels for each Sunday, etc., in the version of the Great Bible (which the Psalter of the Prayer-Book of to-day still follows). Of the Genevan New Testament version there existed a revised edition by Laurence Tomson, which differs slightly from the Genevan translation and contains copious marginal notes. It is frequently bound up with the Genevan Old Testament. Shakespeare may therefore have heard or read *four* different versions of portions of the Scriptures. The bible he would have been most likely to use himself, was the Genevan Version, as we might infer on *a priori* grounds. This, like the Bishops' Bible, contained the Apocrypha, though they are wanting in some copies.

The Bible-Versions I used are the following:—

1. 'The Holy Byble, conteynyng the Olde Testament, and the New. Set 'foorth by aucthoritie. *1576*.'—The colophon states that it was 'Im'printed at London, by Richarde Iugge'. This is the *Bishops' Bible.* (It is a quarto volume in black-letter.)
2. [Title page wanting]. A copy of the *Geneva Bible* with this colophon: 'Imprinted at London, by Christopher Barker Printer to the Queenes 'Maiestie. *1578*.' (Folio in black-letter.)
3. 'The Bible, that is the Holy Scriptures conteined in the Old and New 'Testament . . . London . . . Christopher Barker . . . *1599*.' The Title-page before the New Testament is this: 'The New Testament of 'our Lord Jesus Christ, Translated out of Greeke by Theod. Beza: 'With briefe Summaries and expositions upon the hard places by the 'said Authour, Ioac. Camer. and P. Loscler. Villerius. Engelished by 'L. Tomson . . . London . . . Christopher Barker . . . 1599.' A quarto volume in beautiful Roman type, with an appendix of metrical Psalms. The Apocrypha are wanting. This is the *Geneva-Tomson Bible.* The text of the New Testament differs in a few places from the text of 2. The Old Testament is the same in both. When I cite the text of the Genevan Version I usually quote from the copy of 1578, with which the 1599 copy is to be understood to agree unless otherwise stated.

4. I have compared the New Testament of the "Great Bible", *1539*, as given in the "English Hexapla". London, 1841, and the 'Epistles, and 'Gospels, to be used at the celebration of the Lords Supper', in 'The 'Booke of Common prayer', etc. London . . . Christ. Barker, *1586*. I compared, too, the Prayer-Book, Anno *1559*, reprinted in the Parker Society Publications.
5. The Great Bible of 1549, entitled 'The Byble in Englishe, that is, the 'olde and new Testament, after the translacion appoynted to bee read 'in the Churches'. (Brit. Mus.).
6. The Psalms of 1. are not in the version proper to the Bishops' Bible. These I quote from the Bishops' Version of 1572.
7. I have also compared several other copies of Elizabethan Bibles.

THE ELIZABETHAN BIBLE-TEXTS DIFFER FROM THE TEXT OF THE AUTHORISED VERSION

and from one another. The following examples will show that Shakespeare's text, differing from our Authorised Version, agrees with the Elizabethan.

I) "this dishonour in thine age
"Will bring thy HEAD with sorrow to the ground!"
(2. Henry VI., Act II, III, 19.)

This is an allusion to Genesis, XLII, 38, where our Authorised Version reads:

then shall ye bring *down* my gray HAIRS with sorrow to the grave.

The version of the Genevan, the Bishops', and the Great Bible is:

ye shall bring my gray HEAD with sorow unto the grave.

II) "SABA was never
"More covetous of wisdom and fair virtue
"Than this pure soul shall be." (Henry VIII., Act V, v, 24.)

See 1. Kings, X, where the Authorised Version has 'SHEBA', The Bishops' Bible and the 'Great Bible': 'SABA'. The Geneva: 'Sheba'; but in the heading of the chapter: 'Saba'.

III) "SATAN, AVOID! I charge thee, tempt me not."
(Com. of Err., IV, III, 48.)

[Comp. "Avoid then, fiend!" (*ib.*, v. 66); "false fiend, avoid!"
(2. Henry VI., I, IV, 43)].

Here there is a plain allusion to the temptation-story, see Matth. IV, 10, where the Authorised Version has: GET THEE

HENCE, SATAN; the Bish. Bible: 'Geat thee hence behinde 'me Satan'. It must have been the text given by the Geneva Bible and by the Great Bible (whose version was in the Prayer-Book, Gospel for the first Sunday in Lent) which Shakespeare had in mind:

Then said Iesus unto him, AVOYDE SATAN.[1]

IV) "To pray for them that have done SCATHE to us."
(Rich. III., Act I, III, 317.)

Compare the Authorised Version, Matth., V, 44:

pray for them which DESPITEFULLY USE you, and persecute you,

for which the Genevan, the Bishops', and the Great Bible give:

pray for them which HURT you, and persecute you.

This comes nearer to Shakespeare's text.

THAT SHAKESPEARE WAS FAMILIAR WITH THE GENEVA BIBLE, the following passages, compared with the contemporary versions of the Bible, go to show:

I) "the king's *name* is a TOWER of strength."
(Rich. III., Act V, III, 12.)

Compare Prov., XVIII, 10, in the Genevan version:

The name of the Lord is a strong TOWER.

Instead of 'tower' the Bishops' Bible and the Great Bible read:

'CASTLE'.

[The Authorised Version has 'tower'.]

II) "NEBUCHADNEZAR" in the Folio edition of All's Well (IV, v, 21)— and no earlier edition exist—agrees with the form of the name as given by the Geneva Bible. The Bishops' and the Great Bible have *Nabuchodonosor*.

III) "You would think that I had a hundred and fifty tattered *prodigals* "lately come from swine-keeping, from eating draff and HUSKS". (1. Henry IV., Act IV, II, 36.)

---

[1] I quote from the Geneva Bible 1578 with which the version in the Prayer-Book and Great Bible agrees (except that it has 'saith' for 'said').—The temptation-narrative as given by Luke does not help us. For Luke, IV, 8, the 'Great' and the 'Geneva' Bibles have: 'Hence from me, Satan'; the 'Bishops': 'Geat thee hence 'behynde me Satan'; the Authorised Version: 'Get thee behind me, Satan'.

Compare what the Geneva Bible says about the prodigal son, Luke, XV, 16:

> And he would faine have filled his bollie with the HUSKES that the swine ate.

Instead of 'husks' the Bishops' Version and the Great Bible read: 'CODDES'. (Comp., too, 'As You Like It', I, I, 39—42.)

IV) In 2. Henry VI., Act IV, IV, the messenger having related the doings of Jack Cade and his men, the king exclaims:

> "O graceless men! they KNOW not what they do".

This is, of course, a quotation from Luke, XXIII, 34. Instead of 'KNOW' of the Geneva Bible, the Bishops' Version and the Great Bible have 'WOTE'.

V) In the Merchant of Venice (I, III) the strategy of the patriarch Jacob is instanced in the passage:

> "That all the eanlings which were streak'd and pied
> "Should fall as Jacob's hire . . . .
> "The skilful shepherd peel'd me certain wands
> "And . . . . stuck them up before the fulsome ewes,
> "Who then conceiving did in eaning time
> "Fall PARTI-COLOUR'D *lambs*, and those were Jacob's."

As Halliwell-Phillipps[1] pointed out, the word parti-coloured must be due to the Geneva version, Genesis, XXX, 39, 40:—

> [They] brought forth yong of PARTIE COLOUR, and with small and great spots. . . . . . these *lambes* PARTIE COLOURED . . . .

The Bishops' Bible reads:

> [They] brought foorth lambes RYNGSTRAKED, spotted, and PARTIE . . . . . these RYNGSTRAKED, etc.

The Great Bible (XXX, F), also, has the word 'party coloured'. But it is a much less likely source.

[1] 'An attempt to discover which version of the Bible was that ordinarily used by Shakespeare', 1867. As only ten copies of this booklet were printed Halliwell-Phillipps evidently did not think highly of his own production. I have not seen it. Even the British Museum does not possess a copy. I only know the book through Carter's '*Shakespeare Puritan and Recusant*' (a *Tendenz* book), 1897, p. 197. The conclusion Halliwell-Phillipps arrives at is, that the poet used the Genevan version.

THE BIBLE USED IN THE CHURCH.

Shakespeare's acquaintance, too, with the Bible text used in the church the following passages compared with the different versions go to evince:

1) "A horson ACHITOPHEL; a Rascally-yea-forsooth-knave",

2. Henry IV., Act I, II, 41, Folio;—the Quarto gives the *same form* of the name, which is also found in the Bishops' Bible. The present Authorised, the Genevan, and the 'Great Bible' versions have *Ahithophel*.

2) "for wisdom cries out in the streets, and *no man regards* it." (1. Henry IV., Act I, II, 99.)

Compare Prov., I, 20 and 24, in the Bishops' Version:

v. 20: Wisdome crieth without, and putteth foorth her voyce in the streetes.

v. 24: Because I have called, and ye refused, I have stretched out my hande, and *no man regarded.*

Instead of these italicized words the Geneva Bible has: *'none 'woulde regard'*. The Great Bible reads: *'and no man re-'garded it'*.

3) "Did they not sometime cry, 'all HAIL!' to me?
"So Judas did to Christ." (Rich. II., Act IV, I, 169.)
"so Judas kiss'd his master,
"And cried 'all HAIL!' when as he meant all harm".
(3. Henry VI., Act V, VII, 33.)

Compare Matth., XXVI, 49, where the Bishops' Bible reads:

And foorthwith when he came to Iesus, he sayd, HAYLE maister: and kissed hym.

This agrees also with the words of the Great Bible which were included in the Prayer-Book in the Gospel for the Sunday before Easter. The Geneva Bible does not give the word 'hail' except as a marginal reading. 'Hail' is, however, in Tomson's translation of Mark, XIV, 45, where no other Version has it.

THE APOCRYPHA.

We should remember that Shakespeare was familiar with the Apocryphal Books of the Old Testament which no longer find a place in the common editions of the Authorised Bible.

The History of Susanna.

In 'The Merchant of Venice' (Act IV, I, 223) the Jew is made to say:

> A Daniel come to judgement! yea, a Daniel!
> O wise young judge, how I do honour thee!
> (Cp., too, v. 333 & v. 340.)

Here Shakespeare is thought to have had in view 'The historie of Susanna', where 'a yong child, whose name was Daniel'[1] proves wiser than the judges and convicts the two wicked elders 'of false 'witnesse by their owne mouth', thus saving Susanna from imminent death. 'Fro[m] that day forth was Daniel had in great reputation in 'the sight of the people'. But I ought not to forget remarking that, apart from this Susanna incident, Daniel's wisdom was held in high esteem of old. Compare, e. g., what is said about him in Daniel V, 11, 12:

> There is a man in thy kingdom, in whom is ỹ spirit of the holy gods, and in the dayes of thy father, light and understanding and wisdome like the wisdome of the goddes, was found in him . . . . Because a more excellent spirit, and knowledge, and understanding . . . . were found in him, etc.

Comp., too, Daniel, IV, 6; VI, 3; Ezech., XXVIII, 3; The Bel and the Dragon; etc.

The above narrative of Susanna is also apparently alluded to in *All's Well* (II, I, 141):

> He that of greatest works is finisher
> Oft does them by the weakest minister:
> *So holy writ in babes hath judgement shown,*
> *When judges have been babes.*"

The History of Bel and the Dragon.

In Much Ado About Nothing (Act III, III, 143) Borachio says that fashion makes young men appear "sometime like god Bel's "priests in the old church-window",—the subject of the painting being taken from the apocryphal narrative, where Daniel detects the imposture of the priests of Bel.

[1] I am quoting from the Geneva Bible, 1578. 'young child' is also in the Bishops' Bible and in the Great Bible. The Authorised version substituted 'young youth' for 'young child', which latter agrees better with the passage quoted from All's Well (II, I, 141). See Clarend. Press ed. of Merch. of Ven., p. 120.

THE WISDOM OF JESUS THE SON OF SIRACH, CALLED ECCLESIASTICUS.

The saying of Jesus, the son of Sirach, (*Ecclesiasticus*, chapt. XIII, 1)

He that toucheth pitche, shalbe defiled

had no doubt become a current proverb. Shakespeare refers to it in Much Ado, III, III, 60:

they that touch pitch will be defiled;

in 1. Henry IV., Act II, IV, 455:

this pitch, as *ancient writers* do report, doth defile;

and in other dramas (Love's Lab. Lost, IV, III, 3; All's Well, IV, IV, 24; 2. Henry VI., Act II, I, 196; Timon, I, II, 231).

In illustration of Merch. of Ven., IV, I, 184—186:

The quality of mercy is not strain'd,
It droppeth as the gentle rain from heaven
Upon the place beneath.

the commentators quote from *Ecclesiasticus*, XXXV, 20:

Oh, how faire a thing is mercie, in the time of anguish and trouble! It is like a cloude of raine, that commeth in the time of a drought.

Compare, too, *Deuter.*, XXXII, 2.

THE MACCABEES.

Judas Maccabaeus, the hero of the Apocryphal *Maccabees*, appears as one of the Nine Worthies in Love's Lab. Lost, Act V.

FRENCH BIBLE.

In Henry V. (Act III, VII, 68) the Dauphin is made to say:

Le chien est retourne a son propre vemissement est la leuye lauec au bourbier.

This passage as it stands in the 1st Folio contains three obvious misprints: 'vemissement' for 'vomissement', the second 'est' for 'et', and 'leuye' for 'truye'=truie.

The quotation is taken from the French Bible (2. Peter, II, 22), of which there were differing versions in use in the sixteenth century, most of them being apparently re-edited from, or based on, Olivetan and Calvin's Translation. I have compared several French Bibles and New Testaments. The text coming nearest to the above quotation I find given by La Sainte Bible, Lyon MDXXXXX [*sic*] (par Balthazar Arnoullet):

Le chien est retourné à son propre vomissement: & la truye lauee est retournee au bourbier.

This is identical with Shakespeare's wording except for 'est re-'tournee' repeated here with the required change of gender. Le Nouveau Testament à Lyon, 1584, offers the same version as the above Lyon Bible, but spells 'truie' instead of 'truye'. The French Bibles printed at Geneva 1588, 1605, at la Rochelle 1616, and the New Testament, Geneva 1562, all agree in rendering the latter part of the verse thus:

& la truye lauee *est retournee* à se veautrer au bourbier.

(The italics are in the original text.) Other Bibles differ still more.

LATIN BIBLE.

The phrase *Medice, teipsum* used by the Cardinal in 2. Henry VI. (II, I, 53), which is from the Vulgate (Luke, IV, 23: *Medice, cura te ipsum!)* probably enjoyed proverbial currency.

THE BOOK OF COMMON PRAYER REFLECTED IN SHAKESPEARE'S WORKS.

In the present section I purpose to follow up traces and reminiscences of the Prayer Book in the poet's works, a subject which ought to be of sufficient interest to demand the attention of the Shakespearean scholar. The Prayer Book, I may state, has experienced comparatively few changes since the Elizabethan age, so that Shakespeare worshipped according to the same rites, read the same version of the Psalms, and used much the same prayers, etc., as the English churchman of to-day. I quote from 'The Booke of Common prayer', 1586.

I begin with the phrase: *'world without end'* used by the poet in Love's Lab. Lost (V, II, 799) and in Sonnet 57,5, and no doubt caught from the Prayer Book, where it occurs so frequently.

In

'THE LETANIE'

the words *Good Lord deliver us* form the recurring response to such supplications as, 'From all evill and mischiefe, from sinne', etc., 'From 'fornication, and all other deadly sinne, and from all the deceits of 'the world, the flesh and the devill'.

Compare Tam. of the Shrew, I, I, 66:

Hor. From all such devils, good Lord deliver us!
Gre. And me too, good Lord!

BAPTISM.

The Priest administering baptism puts the following question:

Doest thou forsake the devil and all his workes, the vaine pompe and glory of the worlde, with all covetous desires of the same, etc.?

Compare,

Vain pomp and glory of this world, I hate ye.
(Henry, VIII., Act III, II, 365).

What King Henry V. says in the drama called after him (Act I, II, 31):

what you speak is in your conscience wash'd
As pure as sin with baptism—

agrees with the belief expressed in the baptism service:

Almightie and everlasting God, which . . . . by the Baptisme of thy welbeloved Sonne Iesus Christ, diddest sanctifie the flood Iordan and al other waters, to the mysticall washing away of sinne.
(Comp. Acts, XXII, 16).

THE CATECHISM.

'A Catechisme, that is to say, an Instruction to be learned of every 'childe, before he be brought to be confirmed of the Bishop'. From *à priori* considerations we might infer that young Will Shakespeare learned THE TEN COMMANDMENTS from the Catechism, and not from Exodus, XX, or Deuteronomy, V. This is established *à posteriori* by a passage in Richard III., Act I, IV, 200—202:

the great King of kings
Hath in the *tables of his law* commanded
That *thou shalt do no murder.*

This agrees with the words of the Catechism: 'Thou shalt doe no murther', for which the Geneva Bible, the Great Bible, and the Bishops' Version, and the Authorised Version have: 'Thou shalt not kill' (or Do not kill). See Exod. XX, 13; Deut. V, 17; Matth. V, 21; Rom. XIII, 9; Mark X, 19; Luke XVIII, 20; Jam. II, 11.—Only for Matth. XIX, 18, the Bishops' Bible and the Authorised Version give the same wording as the Catechism.

The Decalogue is moreover quoted from, or alluded to, in the following passages:—

Could I come near your beauty with my nails,
I'ld set my ten commandments in your face.
(3. Henry VI., I, III, 144.)

Thy *sins are visited in this poor child;*
The canon of the law is laid on him,

> Being but the second *generation*
> Removed from thy sin-conceiving womb, etc.[1]
> (King John, II, I, 179.)
>
> *The sins of the father are to be laid upon the children.*
> (Merch. of Ven., III, v, 1.)
>
> *so the sins of my mother should be visited upon me.* (v. 15.)
>
> we shall see wilful *adultery* and *murder committed.*
> (Henry V., II, I, 40.)
>
> *Lucio.* Thou concludest like the sanctimonious pirate, that went to sea with the *Ten Commandments,* but scraped one out of *the table.*
> *Sec. Gent.* '*Thou shalt not steal*'? (Meas. f. Meas., I, II, 7.)

The 10th Commandment is alluded to in The Tamg. of the Shrew, III, II, 232:

> She is my goods, my chattels; she is *my house,*
> My household stuff, my field, my barn,
> My horse, *my ox, my ass, my any thing.*

Respecting his duty towards his neighbour the catechumen is made to learn: 'My duetie towards my neighbour is, to love him as my selfe '[cf. Matth., XIX, 19] .... To keepe my hands from picking and 'stealing', etc.

Compare:—Thou hast my *love:* is not that *neighbourly*?
(As You Like It, III, v, 90.)

The phrase '*pickers and stealers*' (Hamlet, III, II, 348) for 'hands' was probably suggested by the words just quoted from the Catechism.

'My good childe', says the catechist, 'know this, that thou art not 'able to do these things of thy selfe, nor to walke in the co[m]mande- 'ments of God, and to serve him, *without his speciall grace*', etc.

Compare:

> For every man with his affects is born,
> Not by might master'd *but by special grace.*
> (Lov. Lab. Lost, I, I, 152.)

THE PATERNOSTER

is echoed in the following passages:

1) Your will be done. (Rich. II., Act I, III, 144);
God's will be done! (2. Henry VI., III, I, 86);
Cp. Tempest, I, I, 71; Cymb., V, I, 16;—etc.

[1] 'I ..... visite the *sinnes* of the fathers upon the children, unto the third 'and fourth generation', etc. (Catech.). Instead of 'sins' the Geneva Bible and the Authorised Version have '*iniquity*' (Exod., XX, 5); The Bishops' Bible and the Great Bible: *sin* (singular).

II) Forgive us our sins! (Othello, II, III, 116);
O, forgive me my sins! (Tempest, III, II, 139);
I as free forgive you
As I would be forgiven: I forgive all.
(Henry VIII., Act II, I, 82.)

Instances might be multiplied.

III) The following passage:

Isab. Heaven keep your honour safe!
Ang. [Aside] Amen:
For I am that way going to temptation,
Where prayers cross. (Meas. f. Meas., II, II, 157)—

no doubt contains a reference to the petition: 'Lead us not into temptation'.

THE FORM OF SOLEMNIZATION OF MATRIMONY.

The English Marriage Service, which has remained unaltered (excepting a few slight changes in phraseology) since Shakespeare's time, is referred to on several occasions. First, what Benedick says in Much Ado About Nothing, Act V, IV, 29,

this day to be *conjoin'd*
In the *state*[1] of *honourable marriage*,

may be an echo of the opening words of the Marriage Ceremony:

Dearely beloved friendes, we are gathered together here in the sight of God, and in the face of his congregation, to *ioyne together* this man and this woman in holy *Matrimonie*, which is *an honourable estate*, etc.

The Priest proceeding says:

Therefore, if any man can shew any iust cause, why they may not lawfully be ioyned together, let him now speake, or els hereafter for ever holde his peace.

And also speaking to the persons that shall be maried, he shall say.

I Require and charge you (as you wil answere at the dreadfull day of iudgement, when the secrets of all hearts shall be disclosed) that if either of you do know any impediment, why ye may not be lawfully ioyned together in Matrimonie, that ye confesse it, etc.[2]

This is alluded to in Much Ado, IV, I, 12:

Friar. If either of you know any inward impediment why you should not be conjoined, I charge you, on your souls, to utter it.

---

[1] Some editors read: *estate*.

[2] An exhortation similar to this is *now* also pronounced at the publication of the banns. Whether it was done in Shakespeare's time, too, I know not. I scarcely think so. The formula is at least *not* printed in the old Prayer Books.

Compare, too, Act III, II, 91:—

D. John [*to Claudio*]. Means your lordship to be married to morrow?
D. Pedro. You know he does.
D. John. I know not that, when he knows what I know.
Claud. If there be any impediment, I pray you discover it.

The Priest, continuing, says unto the man.

N. Wilt thou have this woman to thy wedded wife', etc.?

The man shall answere.

I will.

Then shall the Priest say unto the woman.

N. Wilt thou have this man to thy wedded husband, to live together after Gods ordinance, in the holy estate of Matrimonie? Wilt thou obey him, and serve him, love, honour, and keepe him in sicknes and in health, & forsaking all other, keepe thee onely unto him, so long as you both shall live?

The woman shall answere.

I will.

Then shall the minister say.

Who giveth this woman to be maried to this man?

Then shall the man say.[1]

I N. take thee N. to my wedded wife, to have and to hold from this day forward, for better, for worse, for richer, for poorer, in sicknesse and in health, to love and to cherish, till death us depart, according to Gods holy ordinance: and thereto I plight thee my trouth.'

The bride uses similar words:

I N. take thee N. to my wedded husband, etc.

Compare, first, As You Like It, Act IV, I, 124—140:

Ros. ... Come, sister, you shall be the priest and marry us. *Give me your hand*, Orlando. What do you say, sister?
Orl. Pray thee, marry us.
Cel. I cannot *say the words.*
Ros. You must begin, '*Will you, Orlando—*'
Cel. Go to. *Will you, Orlando, have to wife this Rosalind?*
Orl. I will.
Ros. Ay, but when?
Orl. Why now; as fast as she can marry us.
Ros. Then you must say '*I take thee, Rosalind, for wife.*'

[1] The Book of Common Prayer, prefixed to the Geneva Bible, 1578, states more fully: 'And the Minister receiving the woman at her father or friends hands, shall cause the man to take the woman by the right hande, and so either to give their troth to the other, the man first saying. I. N., *etc.*

Orl. *I take thee, Rosalind, for wife.*
Ros. I might ask you for your commission; but *I do take thee, Orlando, for my husband:* there's a girl goes before the priest.

Compare, too, Much Ado, Act IV, I, 1—11.

The ceremony of *giving away* the bride to the bridegroom is performed in Much Ado, Act IV, I, 25:

Claud. Stand thee by, friar. Father, by your leave:
Will you with free and unconstrained soul
Give me this maid, your daughter?
Leon. As freely, son, as God did give her me.

Compare Act III, v, 59 (immediately before the marriage ceremony of Act IV, I):—

Messenger: My lord, they stay for you *to give your daughter to her husband.*
Leonato: I'll wait upon them: I am ready.

In As You Like It, Act III, III, 68, Sir Oliver Martext, a vicar, who purposes to marry Touchstone and Audrey, asks:

Is there none here to *give the woman?*
Touch. I will not take her on gift of any man.
Sir Oli. Truly, she must be *given*, or the marriage is not lawful.
Jaques [*Advancing*]: Proceed, proceed: *I'll give her.*

The formula spoken by the woman (which is similar to that pronounced by the man, see above), 'I N. take thee N. to my wedded 'husband, to have and to hold from this day forward, for better, for 'worse', etc., is alluded to in Hamlet, Act III, II, 261:—

Oph. You are keen, my lord, you are keen.
Ham. It would cost you a groaning to take off my edge.
Oph. *Still better, and worse.*
Ham. *So you must take your husbands.*

About the ring, the symbol of marriage in all christendom, I need not waste any words.

Then shall the Priest ioyne their right hands together, and say.

Those whom God hath ioyned together, let no man put asunder.

Compare, 3. Henry VI., Act IV, I, 21:

No, God forbid that I should wish them *sever'd*
Whom God hath join'd together; ay, and 'twere pity
To *sunder* them that yoke so well together.

In 'Romeo', Act IV, I, 55, Juliet says to the Friar:

God join'd my heart and Romeo's, *thou our hands.*

What I have just quoted from the Book of Common Prayer is, of course, ultimately derived from Holy Writ, Matthew, XIX, 6, and Mark, X, 9. But the wording of both the Geneva Bible and the Bishops' Bible and, I may add, of the Authorised Version, *differs from that of the Prayer Book.*

The Bishops' Bible has:

> Let not man therefore put asunder, *that which* god hath *coupled* togeather. (Matth., XIX, 6.)
>
> Therfore, *what* God hath *coupled* togeather, let not man separate. (Mark, X,9.)

The rendering of the Geneva, and of the 'Great', Bibles agrees with that of the Bishops' Bible.[1]

The fourth Act of Much Ado opens with the following words spoken by Leonato:

> Come, Friar Francis, be brief; only to the *plain form* of marriage, and you shall recount *their particular duties* afterwards.

The 'particular duties' may refer to the long enunciation of the duties of man and wife at the close of the Marriage Service. Some duties, too, set forth in the form of a question, are contained in the formulae beginning 'Wilt thou', etc. (see above).—Kate's words, at the end of the Taming of the Shrew (Act V, II, 135 seq.), which are calculated to impress on "headstrong women What duty they do owe "their lords and husbands" (v. 130), are thoroughly in keeping with the duties of wives towards their husbands, as declared in the Marriage Service.

THE BURIAL OF THE DEAD.

Burying of the dead takes place twice in Shakespeare's plays (Cymb., IV, II, and Hamlet, V, I). But there are no rites performed specially characteristic of the English Burial Service. Ophelia's remains are denied Christian burial. 'The rubric before the Burial Office forbids 'it to be used for persons who have laid violent hands upon them-'selves'. But Ophelia has not committed suicide according to Act IV, VII.

The Queen, scattering flowers on the corpse, (Hamlet, V, I, 266) says:

> *Sweets to the sweet:* farewell!

after the manner of the Priests who, while the earth shall be cast upon the Body by some standing by, say:

[1] Tomson's Version is, as I expected, identical with the translation given by the Geneva Bible. Tomson's marginal note to 'coupled', Matth. XIX, 6, offers a point of interest in illustration of 3. Henry VI., Act IV, I, 23, quoted above: 'Hath made them *yoke-fellowes*, as the marriage it self is by a borowed kinde 'of speech called *a yoke.*'

. . . , we therefore commit his body to the ground, earth to earth, ashes to ashes, dust to dust . . . .

THE PSALMS.

In the English Church the Psalms are read or chanted daily in the version retained from the Great Bible of 1539. This version, which is imbedded in the Book of Common Prayer and which has remained practically unaltered through all the centuries, differs, as is well known, from all the translations of the Psalter made since that date. To Shakespeare, English churchman as he was (perhaps he had belonged to the choir of Trinity Church at Stratford, when a boy), the Psalms would be MOST FAMILIAR IN THE PRAYER BOOK VERSION. This will be apparent from a comparison with the translations of the Geneva Bible and the Bishops' Bible. Striking instances I have occasionally noted. (See below (2), (5), (6), (8), (9), (10), (12), (20).)

How thoroughly Shakespeare was conversant with the Book of Psalms will be made evident by the following parallelisms. The passages from the Psalms I quote from 'The Booke of Common prayer' of the year 1586.

(1) *Ps. VII, 15:* Beholde, hee travaileth with mischiefe.
v. 17: For his travell[1] shall come upon his owne head: and his wickednes shall fall on his owne pate.—
Cf. *Ps. CXL, 9:* Let the mischiefe of their owne lips fall upon the head of them: that compasse me about.

Compare:

"O God, what mischiefs work the wicked ones,
"Heaping confusion on their own heads thereby!"
(2. Henry VI., Act II, I, 186-7.)

(2) *Ps. XXII, 12, 13:* Many oxen are come about me: fat *bulles of Basan* close me in on every side. They gape upon me with their mouthes: as it were a ramping and *roaring* Lion.
Cf. *Ps. LXVIII, 15:* As the *hill of Basan*, so is Gods hill: even an high hill, as the hill of Basan.

Compare:

"O, that I were
"Upon the hill of Basan, to outroar
"The horned herd! for I have savage cause."
(Ant. and Cleop., III, XIII, 125.)

The Geneva [and the Authorized] Version have *Bashan* instead of *Basan*. The Bishops' Version prefers Basan.

[1] 'mischiefe', Geneva Bible, 1578.

(3) *Ps. XXIV, 7 & 9:* Lift up your heads, O ye gates, and be ye lift up ye everlasting doores: and the king of glory shall come in.

Compare:

"Then, heaven, set ope thy everlasting gates,
"To entertain my vows of thanks and praise!"
(2. Henry VI., Act IV, IX, 13.)

In the *metrical version* of this Psalm we read:

Ye Princes open your gates, stand open the everlasting gate:
For there shall enter in thereby, the king of glorious state.

(4) *Ps. XXV, 6:* Oh remember not the sinnes and offences of my youth.
(See also Job, XX, 11.)

Compare:

"if the sins of your youth are forgiven you, you're
"well to live".
(Wint. Tale, III, III, 124.)

(5) *Ps. XXXIX, 6:* Beholde, thou hast made my daies as it were a spanne long.

Compare:

"A life's but a span" (*Fol:* "Oh, mans life's", etc.)
(Othello, II, III, 74.)

"how brief the life of man
"Runs his erring pilgrimage,
"That the stretching of a span
"Buckles in his sum of age."
(As Y. L. It, Act III, II, 137.)

"Timon is dead, who hath outstretch'd his span."
(Tim. V, III, 3.)

Cp. "My life is spann'd already."
(Henry VIII., Act I, I, 223.)

The Geneva Bible has:

Beholde, thou hast made my daies as an *hand breadth.*

The same is the rendering of the Authorised Version; and the Bishops', too, has:

". . . . as an *hande breadth* long."

(6) *Ps. XLIV, 15*: Thou makest us to be a byworde among the Heathen.

Compare:

"whose cowardice
"Hath made us by-words to our enemies."
(3. Henry VI., Act I, I, 41.)

Geneva Version:

Thou makest us a *proverb* among the natio[n]s.

Instead of 'by-word' in other passages of the Authorised Version the Geneva Bible invariably has: 'a common talke' or simply 'talke', excepting Job, XVII, 6: "He hath also made me a byword of ẙ people." But Shakespeare could have scarcely had this verse in his mind. Nor can Shakespeare be supposed to have had in view Job, *ib.*, or Jeremiah, XXIV, 9, in the versions of the Bishops' and the Great Bibles.

(7) *Ps. LVIII, 4, 5*: They are as venemous as the poyson of a serpent: even like the deafe adder that stoppeth her eares. Which refuseth to heare the voyce of the charmer: charme he never so wisely.

Compare:

"What! art thou, like the adder, waxen deaf?
"Be poisonous too and kill thy forlorn queen."
(2. Henry VI., Act III, ii, 76.)

"for pleasure and revenge
"Have ears more deaf than adders to the voice
"Of any true decision."
(Troil. & Cress., II, ii, 171.)

"my adder's sense
"To critic and to flatterer stopped are."
(Sonn., 112, 10.)

(8) *Ps. LVIII, 6*: let them fall away like water.

Compare:

[they] "fall away Like water."
(Henry VIII., II, ii, 129.)

The Geneva Bible has:

Let them *melt* like waters.

The Authorised:

Let them *melt* away as waters.

The Bishops':

Let them *be dissolved* as into water.

(9) *Ps. LXVIII, 5*: He is a father of the fatherlesse, and defendeth the cause of the widowes: even God in his holy habitation.

Cf. *Ps. CXLVI, 9*: he defendeth the fatherlesse and widowe.

Compare:

"To God, the widow's champion and *defence*."
(Rich. II., Act I, ii, 43.)

The Geneva version gives for Ps. LXVIII, 5:

He is a Father of the fatherlesse, and *a Iudge* of the widowes . . .,

and for Ps. CXLVI, 9:

he *relieveth* the fatherlesse & widowe.—

The Authorised Version gives the same renderings. The Bishops' Version, too, is almost exactly the same as the Genevan Version.

(10) *Ps. LXXX, 13.* Of the vine brought out of Egypt the Psalmist says:

The wilde Bore out of the wood doeth roote it up.

Compare:

"so soon we shall drive back
"Of Alcibiades the approaches wild,
"Who, like a boar too savage, doth root up
"His country's peace."
(Timon, V, I, 166.)

The Geneva Version is:

The wilde bore out of the wood hath *destroyed* it.—

The Authorised Version:

The boar out of the wood doth *waste* it.

The Bishops':

"*rooteth* it up."

(11) In 2. Henry IV., Act III, II, 41, Shallow says:

"death, as the Psalmist saith, is certain to all; all shall die."

The Psalmist nowhere uses these words, though the certainty of death is frequently alluded to. If Shakespeare had any one passage in view it was probably this:

Ps. LXXXIX, 47: What man is hee that liveth, and shall not see death: and shall he deliver his soule from the hand of hell?

Compare, too, Hamlet, I, II, 72:

"All that lives must die."

(12) *Ps. XC, 9*: we bringe our yeeres to an ende, as it were a *tale* that is *tolde.*

Compare Macbeth, V, v, 26, where the poet says:

Life "is a tale
"Told by an idiot."

and King John, III, IV, 108:

"Life is as tedious as a twice-told tale."

Compare Romeo & Juliet: V, III, 230:

"I will be brief, for my short date of breath
"Is not so long as is a tedious tale."

Geneva Bible:

we have spent our yeres *as a thought.*

Bishops' Bible:

we spende our yeeres *as (in speaking) a woorde.*

Authorised Bible:

we spend our years as a tale that is told.

(13) *Ps. XCII, 11*: The righteous shall florish like a Palme tree: & shall spread abroad like a Cedar in Libanus. *Cp., too, Ps. LXXX, 10, 11.*

Compare:

"he shall *flourish*,
"And, *like a mountain cedar, reach his branches*
"To all the plains about him."

(Henry VIII., V, v, 53.)

"love between them *like the palm* might *flourish*"

(Hamlet, V, II, 40.)

"you shall see him *a palm* in Athens again, and
"*flourish* with the highest."

(Timon, V, I, 12.)

(14) *Ps. CIII, 15, 16*: The dayes of man are but as grasse: for he florisheth as a flowre of the field. For as soone as the winde goeth over it, it is gone: and the place thereof shall knowe it no more.

Cf. Job XIV, 1, 2; Is. XL, 6; 1. Pet. I, 24.

Compare:

"How that a life was but a flower
"In spring time."

(As You Like It, V, III, 29.)

(15) *Ps. CIV, 15*: That hee may bring foode out of the earth, and wine that maketh glad the heart of man: and oyle to make him a *cherefull* countenance, and bread to strength mans heart.

Compare:

"Her *vine*, the merry *cheerer of the heart.*"

(Henry V., Act V, II, 41.)

(16) In King Henry V., Act IV, VIII, 110, the English king says:

"O God, *thy arm* was here;
"And *not* to *us*, but *to thy arm alone*,
"Ascribe we all!" etc.—

v. 128. he continues:

"Let there be sung '*Non nobis*' and 'Te Deum'."

Two Psalmes were here before Shakespeare's mind:—

Ps. XLIV, 3, 4: For they gate not the land in possession through their owne sword: neither was it their owne arme that helped them. But thy right hand, and thine arme, and the light of thy cou[n]tenance.

(*Comp. Ps. XCVIII, 2.*)

Ps. CXV: *Non nobis Domine.*

Not unto us, O Lord, not unto us, but unto thy name give the praise.

The idea of 'Non nobis' was caught from Holinshed, who says:

"the king . . . gave thanks to almightie God for so happie a victorie; "causing his prelats and chapleins to sing this psalme: "In exitū Israel de "Aegypto"; and commanded everie man to kneel downe on the ground "at this verse: "Non nobis, Domine, non nobis sed nomini tuo da gloriam." "Which doone, he caused Te Deum, with certeine anthems to be soong; "giving laud and praise to God, without boasting of his owne force or "anie humane power."

(Boswell-Stone, Shakspere's Holinshed, p. 197.)

(17) *Ps. CXVIII, 9*: It is better to trust in the Lord: then to put any confidence in princes.

Cf. *Ps. CXLVI, 2*: O put not your trust in princes, nor in any childe of man: for there is no helpe in them.

Compare:

"O, how wretched
"Is that poor man that hangs on princes' favours!" etc.

(Henry VIII., Act III, ii, 366.)

(18) *Ps. CXIX, 105*: Thy worde is a lanterne unto my feete: and a light unto my paths.

Compare:

"God shall be my hope,
"My stay, my guide and lantern to my feet."

(2. Henry VI., Act II, iii, 24.)

(19) *Ps. CXXXII, 2, 4, 5*: [David] sware unto the Lorde: and vowed a vowe unto the Almightie God of Jacob. I *will not suffer mine eyes to sleepe,* nor mine eye liddes to slumber: neither the temples of mine head to take any rest. *Untill* I finde out a place for the temple of the Lorde: an habitation for the mightie God of Jacob.

Compare:

"Victorious Prince of York,
"*Before* I see thee seated in that throne
"Which now the house of Lancaster usurps,
"I *vow by heaven these eyes shall never close.*"

(3. Henry VI., Act I, i, 21.)

(20) *Ps. CXXXVIII, 5*: If I forget thee, O Hierusalem: let my right hand forget her cunning.

Compare:

"We therefore have great cause of thankfulness;
"And shall forget the office of our hand,

"Sooner than quittance of desert and merit
"According to the weight and worthiness."
(Henry V., Act II, II, 32.)

Geneva Bible:

. . . let my ryght hand forget *to play.*

The Bishops' and the Authorised Versions are identical with the Prayer Book version.

(21) *Ps. CXLIV, 4*: Man is like *a thing of nought.*

Compare Hamlet, IV, II (end):

Ham. . . . the king is a thing—
Guild. 'A thing', my lord?
Ham. Of nothing; etc.[1]

(22) In conclusion, I draw attention to the following passage:

Ps. IV, 8: Thou hast put gladnesse in my heart: since the time that their *corne and wine and oyle* increased.

The phrase 'corn, wine, oil' (with the words in this order) occurs frequently in the Old Testament. Cp. Deut., VII, 13; XII, 17; etc.; 2. Chron., XXXI, 5; Neh., V, 11; etc.

Compare:

"No use of metal, *corn, or wine, or oil.*"
(Tempest, II, I, 153.)

Montaigne, who is Shakespeare's source for the passage, where this verse occurs, has only:

'nul metal; nul usage de *vin ou de bled.*'

Florio's Translation (used by Shakespeare) gives:

'no use of *wine, corne,* or mettle.'[2]

METRICAL PSALMS.

Notwithstanding the example of Germany with its magnificent hymns no congregational hymnody worthy of the name arose in England till Wesley's times. The hymns of the Elizabethan Church could be counted on one's fingers' ends. 'Te Deum', the most famous non-biblical hymn of the Western Church (still forming part of the Morning Service of the Anglican Church, 'We praise thee, O God'), 'Veni Creator', and one or two more, were all the hymns the English Church could boast

1) But cf. Kyd's Spanish Tragedy, III, II, 72f.

2 Compare the following passage in Nash's 'Christ's Tears over Jerusalem', 1593, (Huth Libr., p. 100): "there were . . . . . Store-houses, filled up to the toppe "with victual, *Corn, Wine and Oyle,* sufficient to maintain", etc. Both Shakespeare and Nash knew their Bible.

of.—'Te Deum' is mentioned twice by Shakespeare: Henry V. (Act IV, VIII, 128); and Henry VIII. (Act IV, I, 92). In both cases he follows his source, Holinshed.—In place of hymns proper the Elizabethan Church sung metrical versions of Psalms, for which the Reformed Church of France and Switzerland gave the example. A collection of such versified Psalms set to music (which is said to be German) I find attached to the Prayer-Book before me (1586) and to the Geneva-Tomson-Bible (1599), where the title runs thus:

> 'The Booke of Psalmes: collected into English Meeter, by Thomas 'Sternhold, Iohn Hopkins, and others: conferred with the Hebrew; with apt 'Notes to sing them withall. Set forth and allowed to be sung in all 'Churches, of the people together, before and after Morning and Evening 'Prayer: As also before and after Sermon; and moreover in private houses, 'for their godly solace and comfort, laying apart all ungodly Songs and 'Ballads, which tend onely to the nourishment of vice, and corrupting of 'youth.'

The first words of THE 137TH PSALM versified by W[illiam] W[hittingham] are sung by Sir Hugh Evans, the Welsh parson:

> "When as I sat in Pabylon"

(The Merry Wives, III, I, 24), the air of which is this:—

The music is from Sternhold and Hopkins's versified Psalms, 1586, but with modernised notation which I owe to the kindness of Herr Kawerau, the Organist of the Berlin Dom.

In the Merry Wives (Act II, I, 63) Mrs. Ford says, that Falstaff's disposition and his words

> do no more adhere and keep place together than the Hundredth Psalm to the tune of 'Green Sleeves'.

Every child knows the famous OLD HUNDREDTH, which Shakespeare must have had in mind and which begins thus:

For the Genevan Psalmody the puritans would naturally have a predilection, hence a 'puritan' 'sings psalms to hornpipes' in the Winter's tale (IV, III, 47). The weavers are referred to by Falstaff as psalmsingers, 1. Henry IV. (Act II, IV, 146). Falstaff, by the way, pretends to have lost his voice with 'singing of *anthems*', 2. Henry IV., (Act I, II, 213). Remember, that St. Paul's possessed 'a very fine 'organ, which at evening prayer, accompanied with other instruments, 'is delightful', as Paul Hentzner says, who visited London, while Shakespeare was there. (Ordish, Shakespeare's London, p. 11.)

In conclusion, I add a brief note on

GRACES.

Of Elizabethan 'graces' before and after meat both in prose and in verse specimens may be seen in the Publications of the Parker Society (see index, s. v. grace).

> Gratiarum actiones a cibo [=post cibum] semper concludantur hac precatiuncula: "Deus servet ecclesiam regem vel reginam custodiat . . . . . . et "pacem nobis donet perpetuam. Amen."

So we read in the Preces Privatae, 1564, p. 402.[1]

Compare the Primer[2], 1559: 'Grace after supper', which concludes thus:

> God save our Queen and Realm, and send us peace in Christ. Amen.

---

[1] Parker Society: Private Prayers of the Reign of Queen Elizabeth.
[2] ibid., p. 18.

Some such petition probably formed the conclusion of every grace after meals, though this is not expressly stated in each case. In 'The ABC With the Catechism: That is to say, An *Instruction* to 'be Learned of every Person before he be brought to be Confirmed 'by the Bishop,' London, 1680[3], we find the following prayer added to the graces *before* and *after* meat:

> God save his Church, our King, Queen, and Realm,
> And send us peace in Christ our Lord. Amen.

A grace in prose is said by Timon (Act III, vi, 79—94), in verse by Apemantus (*ibid.* I, ii, 63—72).

In 'Measure for Measure' (Act I, ii, 14 seq.) occurs the following passage:

> *First Gent.* . . . There's not a soldier of us all, that, in the *thanksgiving before*[4] *meat*, do relish *the petition well that prays for peace.*
> *Sec. Gent.* I never heard any soldier dislike it.
> *Lucio.* I believe thee; for I think thou never wast *where grace was said.*
> *Sec. Gent.* No? a dozen times at least.
> *First Gent.* What, *in metre?*
> *Lucio.* In *any proportion* or in any language.
> *First Gent.* I think, or in any religion.
> *Lucio.* Ay, why not? Grace is grace, despite of all controversy: as, for example, thou thyself art a wicked villain, despite of all grace.

Other references to the custom of offering grace at table will be found in: 1. Henry IV., Act I, ii. 19—23; Tam. of Shrew, IV, i, 162; Coriolanus, IV, vii, 3—4; Merchant of Ven., II, ii, 202.

---

[3] I could find no copies, of the ABC with the Catechism, issued by the Anglican Church, between 1549 and 1680 (cp. above, p. 48).

[4] Comp. Camb. (2nd ed.) Edition of Sh., vol. 1, p. 434.

APPENDIX TO CHAPTER 6.

I venture to add here a few bible reminiscences, to which, so far as I am aware, attention has not yet been drawn:

I) *Caliban:* . . . . . "I'll yield him thee *asleep*,
"Where thou mayst *knock a nail into his head.*"

(Temp., III, II, 68.)

This has always reminded me of the horrible assassination of Sisera by Jael. See Judges, IV, 21.

Deborah, another heroine of this chapter, is mentioned in 1. Henry VI., Act I, II, 105.

II) The prophetic words uttered by Archbishop Cranmer regarding the royal infant, Elizabeth, (Henry VIII., Act V, sc. 5) call up to my mind Simeon's prophecies of Christ, see Luke, II.

III) "*Elb.* Marry, sir, by my wife; who, if she had been a woman "cardinally given, might have been accused in *fornication, adultery,* "*and all uncleanliness* there." (Meas. f. M., II, I, 80.)

Comp. Galat. V, 19.

IV) "Withhold thine indignation, mighty heaven,
"And *tempt us not* to bear *above our power!*"

(King John, V, VI, 37.)

Comp. 1. Corinth., X, 13.

V) "Thou god of this great vast, *rebuke these surges.*"

(Pericles, III, I, 1.)

Comp. Matth., VIII, 26; Mark, IV, 39; Luke, VIII, 24.

Other allusions like

"thou scarlet sin" (Henry VIII., III, II, 255);

"If then the tree may be known by the fruit, as the fruit by the tree" (1. Henry IV., Act II, IV, 470)—

etc., etc., are easily recognized by everybody.

CHAPTER 7.

SHAKESPEARE'S EARTH AND HEAVEN.

SECTION 1.—NEW AND STRANGE LANDS AND BOOKS AND TALES ABOUT THEM, AS REFLECTED IN SHAKESPEARE'S DRAMAS.

While considering Shakespeare's relations to the contemporary Oceanic Literature and Books of Travel, I felt that the subject-matter demanded a less superficial treatment than a mere enumeration of his probable and possible sources. The present section thus took shape, while the remaining portion of the chapter suggested itself as a natural complement, though I am conscious that some of the matter goes rather beyond the limits of my theme.

I desire first to take notice of Shakespeare's references and allusions to the new discovered lands, which attracted so vast an amount of attention in Elizabeth's reign, famous for its maritime exploits and enterprizes, which led to the founding of Britain's great colonial Empire and her rule of the waves. For our present purpose it is sufficient to make mere mention of the names of Drake, Cavendish, Frobisher, Davis, and Ralegh, not to forget the Earl of Southampton, the poet's patron, who took an active share in the work of colonization.

This age of maritime discovery and enterprize found an able chronicler and geographer in Richard Hakluyt, whose great work, "The Principall Navigations, Voiages, and Discoveries of the English Nation", appeared in 1589, and, edited anew, in 1598—1600. This compilation contains a great variety of accounts of naval feats and explorations of Englishmen, exclusive for the most part of the deeds of other nations. So that it forms a fit complement to the works of Peter Martyr, Oviedo, Gomara, Ramusio, etc., extracts from whose volumes had been given in an English dress by RICHARD EDEN.[1]

From the lips of travellers and other men, from works like those of Hakluyt, Eden, etc., from tracts and pamphlets, Shakespeare could derive all necessary information relative to the discoveries and the new lands.

MAGELLAN'S CIRCUMNAVIGATION OF THE EARTH.

In 'The Tempest' (I, II, 372) Caliban says of Prospero:

> his art is of such power,
> It would control my dam's god, *Setebos.*

In Act V, I, 261, we find the name Setebos mentioned again.

It was Farmer, who first suggested that Shakespeare had got this name from *The History of Travayle in the West and East Indies, etc.*, 1577, by RICHARD EDEN, the pioneer of British Oceanic Literature and forerunner of the more famous Hakluyt. This work is merely a new and enlarged edition of Eden's *Decades of the newe worlde or west India*, 1555, which 'made to the English public the first really 'collective presentation of the results of the maritime enterprize of 'that time'. This compilation has been re-edited by Edward Arber in his volume "The first Three English books on America" (1885), from which I quote the following passages:

> Departynge frome hense, they sayled to the. 49. degree and a halfe under the pole Antartyke: where beinge wyntered, they were inforced to remayne there for the space of two monethes, all which tyme they sawe no man except that one daye by chaunce they espyed a man of the stature of a giante, who came to the haven daunsyng and syngynge, and shortly after seemed to cast dust over his heade. The capitayne [i. e. MAGELLAN] sente one of his men to the shore with the shyppe boate, who made the lyke signe of peace. The which thynge the giante seinge, was owt of feare and came with the capitaynes servaunte to his presence into a lyttle Ilande. When he sawe the capitayne with certeyne of his coompany abowte hym, he was greatly amased and made signes holdynge uppe his hande to heaven, signifyinge therby that owre men came from thense[2] . . This giante was so bygge, that the heade of one of owr men of a meane stature, came but to his waste. He was of good corporature and well made

[1] The Decades of the Newe Worlde.

[2] cf. Tempest, II, II, 140, *Cal:* "Hast thou not dropp'd from heaven?"

in all the partes of his bodie, with a large vysage paynted, with dyvers coloures, but for the most parte yelowe. Uppon his cheekes were paynted two hartes, and redde circles abowt his eyes. The heare of his headde was coloured whyte, and his apparell was the skynne of a beaste sowde togyther.

[Later on] there came foure other giantes, without any weapons, but had hydde theyr bowes and arrowes in certeyne busshes. The capitayne reteyned two of these which were youngest and beste made. He tooke them by a deceyte in this maner, that gyvynge them knyves, sheares, lookynge glasses, belles, beades of crystall, and such other tryfels, he so fylled theyr handes that they coulde hold no more. Then caused two payre of shackels of iren to bee put on theyr legges, makynge signes that he wold also gyve them those chaynes: which they lyked very wel by cause they were made of bryght and shynynge metal. And wheras they could not cary them bycause theyr handes were full, the other gyantes wolde have caryed them: but the Capitayne wolde not suffer them. When they felte the shakels faste abowte theyr legges, they begunne to doubte: but the Capitayne dyd put them in comforte and badde them stande styll. In fine when they sawe how they were deceaved they rored lyke bulles and cryed uppon theyr greate devyll *Setebos* to helpe them . . .

They say that when any of them dye, there appere. X. or XII. devyls leapynge and daunsynge about the bodye of the deade, and seeme to have theyr boddyes paynted with dyvers colours. And that amonge other, there is one seene bygger then the residue, who maketh great mirth and reioysynge. This greate devyll they caule *Setebos*, and caule the lesse *Cheleule*. One of these giantes which they toke, declared by signes that he had seene devyls with two hornes above theyr heades, with longe heare downe to theyr feete: And that they cast furth fyre at theyr throtes both before and behynde. The Capitayne named these people *Patagoni*.

One of the giants remained some months with Magellan.

On a tyme, as one made a crosse before him and kyssed it, shewynge it unto hym, he suddeynely cryed *Setebos*, and declared by signes that if they made any more crosses, *Setebos* wold enter into his body and make him brust.

(Arber, *ut sup.*, p. 251—2.)

The Patagonian giants may thus be looked upon as the remote ancestors of our Caliban. Malone suggested that Shakespeare took the names of Alonso, Ferdinand, Sebastian, Antonio, Gonzalo, and Francisco from Eden. This is probable, though these and other names of the *dramatis personae* in the Tempest would be fairly common in countries like Italy and Spain.

The passages quoted above are taken from a chapter entitled: 'A briefe Declaration of the Vyage or Navigation made abowte the Worlde. Gathered owt of a large Booke wrytten hereof by Master Antonie Pygafetta Vincentine, Knyght of the Rhodes and one of the coompanye of that vyage in the which, Ferdinando Magalianes a Portugale (whom sum caule Magellanus) was generall Capitayne of the navie'. Shakespeare's knowledge of the story of the first circumnavigation of our globe by Fernão de Magalhães (Magellan), undoubtedly the greatest of ancient and modern navigators, thus seems to be proven.

In 'As You Like It', III, II, 207, Rosalynd is made to say:—

One inch of delay more, is a South-sea of discovery—

which means: The least delay is to me as tediously long, as the wide Pacific to the discoverer. Possibly Shakespeare had in view Magellan's voyage of discovery across this vast expanse of water, who was the first that ever burst into that silent sea.

What the Patagonian word Setebos (which Pigafetta may have possibly misinterpreted) means, I have not been able to make out. Perhaps some American linguist may help us here.[1] Meanwhile it is curious to note that Setebos (or Setibos) is the name of an aboriginal Indian tribe on the heights of Peru.[2] Whether this coincidence is a matter of pure accident or not, I cannot tell.

In conclusion, I add a short bibliographical note. Eden's[3] Narrative of Magellan's voyage is based upon the account to be found in Ramusio's "Navigationi et Viaggi" (vol. 1. 1550). This Italian account had already appeared in print in 1536 in a quarto volume entitled "Il viaggio fatto da gli Spagnivoli atorno a'l mondo", the second part of which gives a mere translation of Jacques Fabre's abridged French version, "Le Voyage et Navigation, faict par les Espaignolz es Isles de Mollucques", etc. (Paris, *circa* 1525), which is based upon an Italian MS. account by Pigafetta. A complete or an original Italian edition of Pigafetta was never published till 1800. Pigafetta's full

[1] See *post*, Addenda and Corrigenda. (A list of old Patagonian words preserved by Pigafetta is to be found in the 52nd vol. published by the Hakluyt Soc., p. 62.)

[2] cf. La Grande Encyclopédie, XXVI, p. 449; Encycl. Brit., XVIII, p. 677.

[3] As to Eden's Life and Labours, see Arber, ut sup., p. XXXVII ff. The article in the Dict. of N. Biogr. is inaccurate.

account has been printed in English in the 52nd vol. published by the Hakluyt Society (1874).[1]

The Setebos-myth, I may add, I have not met with in any other historical or cosmographical work[2] (which makes mention of Magellan's voyage) anterior to the appearance of the Tempest, excepting only in François de Belleforest's edition of Münster's Cosmography (Tome II, p. 2041,—Paris, 1575), Pigafetta being Belleforest's authority.[3]

Arber, in "Transcr. of the Stat. Reg.", vol. V, p. 117 (no. 2667), notes "an account of Fernando de Magalhaens' Voyage round the world, 1519—1522", London, 1580. This book I have not been able to meet with.

CANNIBALS.

South America was long the land of marvels and monsters, the land of lies, to use a modern coinage. Not only did the Empire of the Incas and the horrible feasts of the Cannibals excite the imagination and fancy of the European mind, but the wildest reports regarding giants of a prodigious height, of headless men, of Amazons, of devils visibly tormenting the natives, and last but not least, of El Dorado, the golden city, were spread, and as easily swallowed. Traces of these fables and legends found their way even into some maps of the time, in the shape of miniature drawings and brief notes.

In the mind of a sixteenth century man the word *Cannibal* would be first and principally associated with South America, whence it was

[1] The volume contains six contemporary accounts of Magellan's voyage, Peter Martyr's contemporary version (Decade V, cap. 7) having been omitted, which seems also to have escaped the notice of Winsor (Hist. of America) and Harrisse (Bibl. Am. Vetust.).

[2] In order not to overburden this essay unnecessarily with notes more or less foreign to the subject, I must refrain from quoting the titles of the many works I examined with regard to this point.

[3] Malone (Var. Ed., XV, p. 58) says that "Setebos is also mentioned in Hak- "luyt's Voyages, 1598". This statement is incorrect. Perhaps Malone had in mind Purchas His Pilgrims (1625, Second Booke, p. 35) which he may have confused with Hakluyt's Collection at the moment of writing the note. — I may here also correct another error, to be found in the Encycl. Brit., XVIII, p. 353, where it is asserted that *Settaboth* is the original form of the word in Pigafetta. By what the writer allowed himself to be misled, it is difficult to say. Perhaps he was thinking of a passage in Fletcher's MS.account of Drake's Voyage (printed in Hakl. Soc., no. 16, p. 48), where we find the above spelling of the word.

derived. Other parts of the world might produce man-eaters, but none could beat this quarter of the world for its anthropophagi.

In Othello (I. III, 143), e. g., Shakespeare speaks of

> *the Cannibals* that each other eat,
> *The Anthropophagi.*[1]

Compare Coriolanus, IV, v, 200:

> An he had been *cannibally* given, he
> might have broiled and eaten him too.

(For other instances, see Bartlett's Concordance.)

The word *Caliban* ("a savage and deformed Slave", *Dram. Pers.*) is apparently a variant of Cannibal. Caliban is tormented by spirits, even as the Brazilian barbarians were said to be "wofully even in "this life tormented by the Devill", as De Lery writes, who continues: "For, I have sometimes seene them, even while they were talking "with us, immediately crying out like frantike men, Hei, hei, helpe "us, for Aygnan [the Devil] beateth us. Nay, they affirmed, that the "wicked spirit was seene of them sometimes in the shape of a cruell "Beast, sometimes of a Bird, and sometimes also in some monstrous "forme".[2] (Some editions of De Lery's work contain a pictorial representation of the Devil in the act of maltreating the poor aborigines.) Whether this coincidence is accidental, I know not. It is at

[1] The use of both synonyms 'Cannibals' and 'Anthropophagi' together is fairly frequent in the 16th century writers. E. g., in Richard Eden's 'Treatyse of the newe India', 1553, translated from Münster's Cosmographia', we find a paragraph: "Of the people called Canibales or Anthropophagi, which are accustom- "ed to eate mans fleshe" (Arber: The First Three English Books on America, p. 29). Compare, too, the New Engl. Dict., s v. Cannibal.

[2] De Lery: "Histoire d'un voyage faict en la terre du Brésil". I have given the above translation from "Extracts out of the Historie of John Lerius a Frenchman who lived in Brasill with Mons. Villagagnon, Ann. 1557. and 58" contained in Purchas His Pilgrims (1625), IV, p. 1336. Thevet, in his "Singularites de la France Antarctique", translated into English by Hacket, 1568, has a similar passage: 'these pore men .... are subiect to many fantastical illusions & perse- 'cutions of wicked spirites .... these pore *Americanes* do oftentimes see a wicked 'spirite, sometimes in one forme, & sometimes in an other, the which they name 'in their language *Agnan*, the which spirit persecuteth them day and night, not 'onely their soule, but also their body, beating them, and doing them much in- 'iury, so that you shal hear them make a pitiful cry, saying in their language ... 'seest thou not *Agnan* yt beateth me .... Also the people of *Ginney*, & of *Canada* 'are likewise tormented ....' (The New found worlde, or Antarctike ..., 1568, fol. 52.) Cp. Purchas His Pilgrims, vol. IV, 1267 and 1337.

any rate worth mentioning that such reports were current in the sixteenth century.

Shakespeare, in describing the ideal commonwealth (The Tempest, II, I, 147 seq.) made use of a chapter in MONTAIGNE'S *Essais* (as stated above, s. v. Montaigne, p. 51), which is written *à la* Tacitus in praise of the Cannibals of Brazil or France Antarctique, as the French were pleased to name the country. Montaigne's remarks relate to the native tribes located near Rio de Janeiro, where the French had attempted a settlement under Villegaignon in the year *1555*.

GUIANA AND THE ACEPHALI.

Guiana with its dazzling phantom of El Dorado, we know, lured more than one man to destruction. No less distinguished a man than SIR WALTER RALEGH, whom we honour not only as a great man of letters but as the father of American colonization and promoter of commerce and navigation, "the Shepherd of the Ocean", as Spenser picturesquely calls him, sacrificed his fortune and life to the discovery of the rich gold-mines of Guiana. It was in 1595 that he led his first expedition to this region, an expedition which was followed up by several more within a few years. Immediately on his return, Ralegh published an account of his voyage: *The Discoverie of the large, rich and bewtiful empire of Guiana, with a relation of the Great and Golden City of Manoa (which the Spaniards call El Dorado) .... Performed in the yeare 1595 ... London, 1596.* There were two editions of this year, according to Arber: Transcr. of St. Reg., V, p. 185 no. 4232. This work, an elegant production, as Camden characterized it, attracted such attention that it was translated into the principal European languages. It was reprinted verbatim by Hakluyt in his "Principal Navigations", etc., vol. III, pp. 627 etc. (1600). Ralegh's book was probably read by his contemporaries with as great avidity as we devoured Stanley's "In Darkest Africa". LAWRENCE KEYMIS, too, sent by Sir Walter Ralegh in 1596 to continue his explorations in Guiana, wrote *A Relation of the second voyage to Guiana, Lond. 1596*, which was likewise embodied in Hakluyt's work, vol. III.

To Guiana which roused so much interest and curiosity Shakespeare refers in "The Merry Wives of Windsor", Act I, sc. III, 75, where Falstaff says of Page's wife:

she bears the purse too; she is *a region in Guiana, all gold and bounty.* I will be cheater to them both [i. e. Ford's and Page's wives], and they shall be exchequers to me; they shall be my East and West Indies, and I will trade to them both . . . .

And v. 88: *Falstaff* [to Robin]:—

Hold, sirrah, bear you these letters tightly;
Sail like my pinnace *to these golden shores.*

RALEGH's book *The Discovery of Guiana* contains a marvellous report, which excited a great deal of curiosity in England as well as on the Continent: the legend of Acephali in Guiana.[1] Ralegh, to quote his own words, says:

Next unto Arui there are two rivers Atoica and Caora, and on that branch which is called Caora [a tributary of the Orinoco], are a nation of people, whose heads appeare not above their shoulders; which though it may be thought a meere fable, yet for mine owne part I am resolved it is true, because every childe in the provinces of Arromaia and Canuri affirme the same: they are called Ewaipanoma: they are reported to have their eyes in their shoulders, and their mouthes in the middle of their breasts, and that a long traine of haire groweth backward betweene their shoulders.

(Hakl. Soc., 3, p. *85;* Hakluyt's "Voyages", III, *652*). KEYMIS repeated his fable in his account of 'the second voyage to Guiana' (see Hakluyt's *Princ. Nav.*, III, *677*). It is extremely probable that Shakespeare alludes to these fabulous 'Ewaipanoma' in Othello, I, III, 143:—

And of the Cannibals that each other eat,
The Anthropophagi and *men whose heads*
Do grow beneath their shoulders —

and in The Tempest, III, III, 43:

When we were boys,
Who would believe that there were mountaineers
Dew-lapp'd like bulls, whose throats had hanging at 'em

---

[1] In Southern Venezuela of our times. As to the bounds and limits of Guiana, see Purchas's Pilgrimes, IV, 1270.

Even as late as Humboldt's time these fabulous reports of headless men in South America were repeated in all seriousness (see Hakl. Soc., *3.*, Schomburgk's edition of Ralegh's Disc. of Guiana., p. *85*). Accounts of Acephali had existed since the times of Herodotus and Pliny, and such a nation was written of by Mandeville and Sigismundus ab Herberstein. (See Hakl. Soc., ut sup. p. *85*, and cp. Arber: First Three English Books on America, p. 323, and Hakluyt's Voyages, I, 494.)

Wallets of flesh?[1] or that there were *such men*
Whose heads stood in their breasts? Which NOW *we find*
Each putter-out of five for one will bring us
Good warrant of.

Each putter-out of five for one = each traveller, who, before setting out on a journey, invested his money at a rate of five for one.

ENGLAND'S FIRST COLONIES.

Sir Walter Ralegh's colonial enterprizes having ended in failure, the efforts for the colonization of Virginia were renewed in King James's reign, who issued a charter in 1606, under which England's first permanent colony was established. To this settlement of Virginia Shakespeare probably refers in Henry VIII., Act V, sc. v, 51–53, where Cranmer is made to prophesy of James I:

Wherever the bright sun of heaven shall shine,
His honour and the greatness of his name
Shall be, and make new nations.

The year 1609 proved memorable in the annals of Virginia. James having granted a new charter, a fleet of nine vessels set sail for the new colony at the close of May with the hopes of all England on board. A violent storm, however, scattered the fleet and the admiral-ship, severed from the rest, was driven on the much dreaded Bermuda[2] Islands, which, contrary to expectation, afforded a safe shelter to the shipwrecked persons.

It is probable that Shakespeare had these adventures in his mind when writing his 'Tempest'. The mention of the *still-vex'd Bermoothes* (Tempest, I, II, 229) ought to be emphasized in this connexion.—Too much, however, has perhaps been made of the coincidences.

Shakespeare may have got information about these events from contemporary tracts, of which the best known are (I) Silvester Jourdan's

[1] The goitre, found to exist in the Alps, in mountainous regions of South America, and elsewhere, is well known to be no traveller's tale. Respecting this disease the following passage to be found in Bacon's works (ed. Ellis and Spedding, vol. II, 472) is worth quoting:—"Snow-water is held unwholesome; insomuch "as the people that dwell at the foot of the snow-mountains, or otherwise upon "the ascent, (especially women,) by drinking of snow-water, have *great bags hang-* "*ing under their throats*". (Bacon's theory cannot be discussed here.)

[2] Ralegh, in his Discovery of Guiana, pr. 1596, says: "The rest of the "Indies for calmes, and diseases [are] very troublesome, and the Bermudas a "hellish sea for thunder, lightning, and stormes." (Hakluyt Soc., 3, p. 114.) Compare also the note in the Clarendon Press ed. of The Tempest, p. 89.

"*A discovery of the Barmudas, otherwise called the Ile of Divels*", (II) and "*A true declaration .... published by .... the Councell of Virginia*", both of which appeared in 1610. He must, of course, too, have been told the news, which created vast public excitement, by word of mouth.[1]

It is not only to the Bermudas, however, that Shakespeare refers in The Tempest, but the drama is replete with incidental allusions to *the new lands* and the topics brought up by colonies and colonisation.[2]

ARCTIC VOYAGES.

About the middle of the sixteenth century the English had attempted to discover a north-eastern passage to the East Indies,—an undertaking which ended in failure, but which led to an important trade with Russia in the White Sea. References to this country occur in several plays (Meas. f. Meas, Wint. Tale, Henry V., Macbeth) but especially in 'Love's Labour's Lost', which contains distinct allusions to a contemporary incident relating to Russian-English connexions. (Compare Mr. S. Lee's paper, "A New Study of Love's Labour's Lost", Gent. Mag., Oct., 1880.)

The Dutch, the powerful rivals of the English in the trade with Russia in the White Sea since *1578*), still persuaded that a north-east passage existed and only required to be discovered (which opinion was correct after all, as Nordenskiöld has proved), fitted out three expeditions in three successive years, the most memorable of which was the third (1596—'97) in which Willem Barendsz was pilot. Barendsz and his companions were forced to spend a terrible winter in the frozen latitudes, the first arctic winter that was successfully faced by European navigators. To this voyage, it is natural to suppose with Mr. Coote[3], Shakespeare refers in Twelfth Night, Act III, sc. II, 27:—

> you are now *sailed into the north* of my lady's opinion; *where you will hang like an icicle on a Dutchman's beard.*

The antithetical idea being, of course, the equatorial region of the lady's opinion. With regard to Barendsz's voyage W. A. Wright says:

[1] For fuller information on all this, see Malone XV, 385; and my remarks in the Archiv für das Studium der neueren Sprachen, vol. 107, pp. 170 seq.

[2] cf. Lloyd's "Essays", 1894. p. 2.

[3] Hakl. Soc., 59, p. XCIII; New Sh. Soc. Trans., 1877—"79, p. 94.

. . . . 'A translation of Gerrit de Veer's account of this voyage was 'entered on the books of the Stationers' Company to John Wolfe on 'the 13th of June, *1598*, but the reprint of Phillip's translation for 'the Hakluyt Society [No. 54] is taken from a copy of 1609, and 'apparently an earlier edition is [*not*] known. Shakespeare, however, 'may very well have heard of the voyage before 1602, the date of 'Twelfth Night.' (Clar. Press ed. of Twelfth Night, p. 138.)

To LAPLAND there is a reference in Com. of Err. IV, III, 11.

THE INDIES.

In order to know and remember what these words would mean to an Englishman living about 1600, it is desirable to recall to mind one or two prominent facts. First, it is necessary to bear in mind that the terms had then a wider meaning than they have to-day. Indies or 'India is now applied to all farre-distant countries, not in 'the extreme limits of Asia alone, but even to whole America', so we read in Purchas 'his Pilgrimage' (London, 1613, p. 381). America was often called West Indies in contradistinction to the East Indies of Asia. Of both these Indies Spain was now no longer the unchallenged possessor. The Invincible Armada having found its Moscow in the British waters in 1588, the Indies weere looked upon by the Dutch and English as their proper spoil. Not only were the rich treasure-ships from the Indies openly seized, but East India trading companies were formed both in England and in Holland. In *1591* the first English voyage to the East Indies was undertaken, and on the 31st of December, *1600*, the charter of incorporation was granted by Queen Elizabeth to the 'governor and company of the merchants 'of London trading into the East Indies'.

The Indies are frequently mentioned in Shakespeare's works. I cite the following examples:

Com. of Err., III, II, 136:

> *Ant. S.* Where America, the Indies?
> *Dro. S.* Oh, sir, upon her nose, all o'er embellished with rubies, carbuncles, sapphires, declining their rich aspect to the hot breath of Spain; who sent whole armadoes of caracks to be ballast at her nose.

Compare Othello, I, II, 50:

> Faith, he to-night hath boarded a land carack:
> If it prove lawful prize, he's made for ever.

Again,

they shall be my East and West Indies. (See above, p. 229.)

Again,

from East to western Ind (As You Like It, III, II, 93.);
savages and men of Ind. (Tempest, II, II, 61.)

The words Indies, India, Indian occur *passim.*

To the sun-worshippers either of East India or, as I think, more probably of the New World, we find references in Love's Labour's Lost, IV, III, 221:

Who sees the heavenly Rosaline,
That, like a rude and savage man of Inde,
At the first opening of the gorgeous east,
Bows not his vassal head and strucken blind
Kisses the base ground with obedient breast?—

in "All's Well", I, III, 210:

thus, Indian-like,
Religious in mine error, I adore
The sun, that looks upon his worshipper,
But knows of him no more.—

and in Love's Lab. Lost., V, II, 202:

Vouchsafe to show the sunshine of your face,
That we, like savages, may worship it.

MEXICO

is mentioned in 'The Merchant of Venice', I, III, 19.

ALEPPO, AND "THE TIGER".

Her husband's to Aleppo gone, master o' the Tiger.
(Macbeth, I, III, 7.)

. . . in Aleppo once,
Where a malignant and a turban'd Turk
Beat a Venetian and traduced the state,
I took by the throat the circumcised dog,
And smote him, thus.
(Othello, V, II, 352.)

Orsino, this is that Antonio
That took the Phoenix and her fraught from Candy;
And this is he that did *the Tiger* board,
When your young nephew Titus lost his leg.
(Twelfth Night, V, I, 65.)

Aleppo, "now chiefe Citie of Syria", as Purchas writes,[1] was an important emporium and a meeting-place of Asiatic caravans. The English Government was represented there by an English consul. "The state and trade of [Aleppo]", wrote one Wrag about 1597[2], "because it is *so well knowen* to most of our nation, I omitte to "write of".

It is worthy of remark that an English vessel (or *possibly* two or three different ships) bearing the name *Tiger* actually plied in the Levant, and is traceable in these waters in the years 1583, 1587, and 1628. In Hakluyt's "Voyages", II (1599), we read:

> John Eldred's narrative.—I Departed out of London in the ship called the *Tiger*[3], in the company of Master John Newbery, Master Ralph Fitch and six or seven other honest merchants, on Shrove Monday [12 February] 1583; and arrived at Tripolis of Syria, the 1st day of May next ensuing ...... We passed forward with camels ..... until we came to Aleppo: where we arrived the 21 st of May. This is the greatest place of traffic, for a dry town [i. e., an inland town, not on a great river] that there is in all these parts.

(I am quoting from Arber's reprint, in "An English Garner", 3, pp. 159 and 161.) Linschoten, referring to the same English expedition[4], remarks:

> Tripolis, a town and haven lying on the sea-coast of Syria, where all the ships discharge their wares and merchandise, which from thence are carried by land to Aleppo, which is a nine-days' journey.—In Aleppo, there are resident divers merchants and factors of all nations, as Italians, Frenchmen, Englishmen, Armenians, Turks, and Moors.

(Arber, *ut sup.*, p. 188.)

We find 'the Tiger' again in the Levant in 1587 and 1628. Possibly it is the same vessel referred to above. But the name was rather common. There were at least three different English vessels of different tonnages bearing this name in Shakespeare's days, not to speak of Spanish and Dutch 'Tigers'[5].

[1] Purchas, His Pilgrimage, 1613, p. 74.

[2] Hakluyt's Princ. Navig., etc., vol. II, part. 1, p. 309. cf. also pp. 172, 276, etc. ibid.

[3] "a ship of London, called the *Tiger*" (Arber, Engl. Garner, 3, p. 169).

[4] For a third account of this voyage, written by Ralph Fitch, cp. Arber, ut sup., pp. 167 seq.

[5] I must content myself with bare references here: Clarendon Press ed. of Macbeth, p. 82; Purchas His Pilgrimes, 1625, Second Part, p. 1618; Hakluyt Soc.,

THE NILE, WITH A NOTE ON ABIOGENESIS.

In 'Antony and Cleopatra' (II, vii, 20—'6) Antony is made to say:—

Thus do they, sir: they take the flow o' the Nile
By certain scales i' the pyramid; they know,
By the height, the lowness, or the mean, if dearth
Or foison follow: the higher Nilus swells,
The more it promises: as it ebbs, the seedsman
Upon the slime and ooze scatters his grain,
And shortly comes to harvest.

Passages to illustrate the above verses have been quoted by Reed and Malone from Pliny's 'Natural History', translated by Philemon Holland, 1601, Bk. V, ch. 9, and from Leo's History of Africa, translated by John Pory, 1600, Bk. VIII. (See Var. Ed., XII, p. 263.)[1]

A passage in Latin coming near to Leo's wording is to be found in De Jode's Atlas[2] (Speculum Orbis Terrae, Antwerp, 1593, see 'Africa'), which passage, it is curious to say, is immediately preceded by the following one, which throws light on Lepidus's remarks regarding the abiogenetic origin of serpents and crocodiles. Respecting the Nile we read:

> Hanc igitur portionem [quae utrinque eius ripis adiacet] aestivo sydere exundans limo secum adducto irrigat ac foecundat, adéo efficacibus aquis ad generandum, alendumque, ut praeter id quod scatet piscibus, quod Hyppothamos [*sic*] Crocodilosque vastas beluas gignit: glebis etiam infundat animas, ex ipsaque humo vitalia effingit. Hoc eo manifestum est quod ubi sedavit diluvia, ac se sibi reddidit per humentes campos, quaedam nondum perfecta animalia, sed tum primum accipientia spiritum, & ex parte iam formata, ex parte adhuc terrea versentur. Qua re adducti nonnulli in Aegypto, tum hominem, tum caetera animantia primo ex limo solaribus radiis formatum asseverate ausi sunt: de qua re lege Theodorum Siculum lib. 1. rerum antiquarum cap. 1.

Compare Ant. and Cleop., *ut sup.*, 27—31:

Lep. You've strange serpents there.
Ant. Ay, Lepidus.
Lep. Your serpent of Egypt is bred now of your mud by the operation of your sun: so is your crocodile—

59, pp. LXXII and 157; Arber's English Garner, vol. V, p. 88; Purchas, *ut sup.*, IV, p. 1144; De Jonge 'Opkomst', etc., I, 33; Hulsius's 23rd. Voyage, p. 46; Lefroy, Memorials of the Bermudas, index.

[1] Pory's Leo was extracted by Purchas, in 'His Pilgrimes', 1626, vol. II. As to the Nilometer, see p. 838, and cp. p. 897.

[2] I do not say it was Shakespeare's source.

and also Act I, III, 68:—

Ant. . . . . . . By the fire
That quickens Nilus' slime.

I may remark that the belief in equivocal generation seems to have been general in Shakespeare's days. Compare what Sir Francis Bacon says in his Natural History:

"The insecta are found to breed out of several matters: some breed of "mud or dung; as the earth-worms, eels, snakes, &c".—"intending by "[insecta] creatures bred of putrefaction." (Bacon, ed. Ellis and Spedding, II, p. 557; cp., too, p. 638.)

Perhaps there is a reference to this doctrine in Othello, IV, I, 256—7:

If that the earth could teem with woman's tears,
Each drop she falls would prove a crocodile.

PRESTER JOHN, THE GREAT CHAM, AND THE PIGMIES.

In Much Ado About Nothing, Act. II, i, 271 ff., Benedick is made to say:—

Will your grace command me any service to the world's end? I will go on the slightest errand now to the Antipodes that you can devise to send me on; I will fetch you a toothpicker now from the furthest inch of Asia, bring you the length of *Prester John's* foot, fetch you a hair off *the great Cham's* beard, do you any embassage to *the Pigmies*, rather than hold three words' conference with this harpy.

PRESTER JOHN, who had occupied so prominent a place in the mind of Europe, was originally supposed to be a potentate of enormous power and splendour in the far East, but later he was assigned a locus in Abyssinia or Ethiopia. Robert Hues, a distinguished contemporary of Shakespeare's, says in his *Tractatus de globis* (1638; the first edition, in Latin, appeared in 1593):—

Next to these is the spacious territory of the King of the Æthiopians (who is called Pretegiani, and corruptly Prester John), which kingdome is famous for the long continuance of the Christian Religion in it, which hath been kept amongst them in a continuall succession ever since the Apostles time. These Christians are called Abyssines, but more rightly Habassines . . . . . Their dominion was anciently extended very farre through Asia also.[1]

[1] Hakluyt Soc., 79, p. 78. On Prester John see Encycl. Brit., XIX, p. 714. For further particulars cf. Zarncke, 'Der Priester Johann'; and Hakl. Soc., ser. 2, vol. 10, 1903.

The Great Cham. This appellation was applied to the supreme ruler of a region in the East and North-East of Asia, called Cathay, Tartary, or China, of which Shakespeare's contemporaries had a very hazy idea.[1]

The Pigmies appear in Homer as a small folk in the far southern land by the streams of Oceanus, on whom the cranes made war. Aristotle held that the pigmies inhabited the marshes out of which he supposed the Nile flowed. He was not far wrong. For Sir Henry M. Stanley found them in Central Africa in the great forest. Other writers localized them in different parts of the world. Mercator, in his maps, placed them near the North Pole. For instance, in his Atlas, 1602, we find this legend on an island near the North Pole:—

Pygmęi hic habitant 4 ad summum pedes longi, quem admodum illi quos in Gronlandia Screlingers vocant.

A FABULOUS REPORT.

For Ariel's mimicry of Trinculo (Temp. III, ii) and the noises of the inchanted isle, Shakespeare is supposed to have made use of fabulous reports told by Marco Polo (and repeated by later writers) about the desert near Lop in north-eastern Asia, where devils counterfeited the voices of human beings, and noises of musical instruments were heard in the air.[2] But this is uncertain.

SECTION 2.—ASTRONOMICAL AND ASTROLOGICAL LORE

or Stars and Influence of Stars in Shakespeare's works.

In Shakespeare's days the

PTOLEMAIC SYSTEM

was still the generally accepted and prevalent astronomical theory.

[1] About Cathay cf. Hakl. Soc., vol. 36—37.

[2] Malone, Var. Ed., XV, p. 120. Comp. Arber, 'The First three Engl. Books on America', p. 27.

For completeness' sake only, I subjoin what other names of modern non-European geography are found in Shakespeare's text:—*Arabia, -n; Cataian,* in the sense of thief, sharper; 'frosty *Caucasus*'; *Hyrcania*; *Persia, -n,* with the Sophy=Shah; *Tartar;* 'high *Taurus*' snow'.—*Afric, Africa, -an; Argier* (=Algiers); *Barbary; Canary; Egypt, -ian; Ethiope, Ethiopian; Morocco; Tripoli, -s; Tunis*—'this Tunis, 'sir, was Carthage'.

Though Copernicus had put forward a new explanation, the heliocentric theory, it was looked upon as nothing more than a curious and newfangled doctrine, at which scientists shrugged their shoulders. And it was only after Shakespeare's time that a change was brought about by Galileo, Kepler, and Newton, notwithstanding whose demonstrations, however, the old system remained in vogue till about the end of the seventeenth century.

In order to give a succinct account of this system and an idea of the form in which it was popularly taught in schools and universities all over Europe, till it was superseded by that of Copernicus, I here transcribe the following passage, translated by Dr. Masson (ed. of Milton, 1874, vol. I, p. 90) from a Latin manual or Catechism of Astronomy by Michael Moestlinus, Professor of Mathematics in the University of Heidelberg, and preceptor of Kepler, a younger contemporary of our poet (Epitome Astronomiae, &c., 1582, pp. 34, 35):— "*Quest.* How many are the Orbs, or celestial Spheres, and what is "their order?" *Ans.* "There are various opinions concerning the number "and order of the celestial Spheres; but, following for the present, "for the sake of learners, the doctrine of the Alphonsines, we reckon ten, "in this order:—The 1st is the Sphere of the Moon, which has the "lowest place in Æther; the 2nd that of Mercury; the 3rd that of "Venus; the 4th that of the Sun; the 5th that of Mars; the 6th "that of Jupiter; the 7th that of Saturn. And these are the Spheres "of the Seven Planets, or wandering stars, each of which has only "one star, viz. its own planet, inserted in it. To these an 8th suc-"ceeds, which, from its order, is called 'the Eighth Sphere' but also "'the Firmament', on account of its containing, and as it were fortifying "or walling round, all the other Spheres—for it was believed by the "ancients to be the last and Supreme Sphere. It is also called the "Sphere of the Fixed Stars, because in it are all the rest of the stars, "whatever their number, after the planets are excepted. There is "moreover a 9th, and finally a 10th Sphere; which last is the "Primum Mobile, or Last Heaven. These two Spheres are destitute "of stars".[1] Blundevile,[2] who has a similar passage, adds an 11th Sphere, which had been invented to satisfy theological demands,

[1] For further contemporary authority on the point, read "Batman uppon Bartholome, his Booke De Proprietatibus Rerum" (1582), extracts from which are printed in New. Sh. Soc. Trans., 1877—'79, pp. 436 seq.

[2] His Exercises, 4th ed., 1613, p. 281.

"The Imperial heaven, where God and his Angels are said to dwell".—These concentric hollow Spheres enclosed the stationary Earth, the centre of the Universe, at different distances, 'covering one another 'like the scales of an Onion', as Blundevile has it. It is to be added that the Atmosphere or Air enveloping the Earth and Water was supposed to be surrounded by a region of Fire, below the sphere of the Moon.[1] This in general outlines was the astronomical belief held by men of Shakespeare's palmy days. At the same time, it is to be remembered, that in the infant time of science astronomy and

ASTROLOGY

went hand in hand. Shakespeare always uses the words 'astronomy', 'astronomer', 'astronomical' in the sense of astrology, -er, -ical. 'In-'fluence' in his text everywhere=starry influence (excepting Sonn., 78, 10, and Lov. Lab. L., V, II, 869, where it means inspiration). References and allusions to stars as excercising influence on human and terrestrial affairs are very frequent in Shakespeare. Compare Schmidt's lexicon, *s. v.* 'influence', 'star', 'aspect', etc. The following quotations will serve as examples:

> D. *Pedro* . . . You were born in a merry hour.
> *Beatr.* No, sure, my lord, my mother cried; but then there was a star danced, and under that was I born.
> (Much Ado, II, I, 347.)
>
> In my stars I am above thee.
> (Twelfth Night, II, v, 155.)
>
> My stars shine darkly over me.
> (Twelfth Night, II, I, 3.)
>
> By each particular star in heaven and
> By all their influences.
> (Wint. Tale, I, II, 425.)

Besides the above instances, which might easily be multiplied, others are referred to below.

Did Shakespeare believe in starry influence? We hope not. But when we see an enlightened man of science, like Bacon, unable to free himself completely from astrological notions,[2] we may suppose that the great dramatist would have given a reserved answer to any enquirer desirous of knowing his opinion on this point. However, he certainly regarded *much* of the old faith (if not all of it) to be mere skimble-skamble stuff, as a passage quoted below the text clearly shows.[3]

[1] An explanation that had been given by Aristotle.

[2] See Bacon, ed. by Ellis and Spedding, vol. IV, p. 349 ff.

[3] "This is the excellent foppery of the world, that, when we are sick in fortune,—often the surfeit of our own behaviour,—we make guilty of our disasters

To return to our subject, WHAT TRACES of the above astronomical system do we find in Shakespeare's text?

AN ILLUSTRATIVE PASSAGE FROM TROILUS AND CRESSIDA.

First, the following passage in 'Troilus and Cressida', I, III, 85ff., deserves to be quoted at length:

"The *heavens*[1] themselves, the *planets* and *this centre*[2]
"Observe degree, priority and place,
"Insisture, course, proportion, season, form,
"Office and custom, in all line of order;
"And therefore is the glorious *planet Sol*
"In noble eminence enthroned and *sphered*
"*Amidst the other*;[3] whose medicinable eye
"Corrects the ill *aspects of planets* evil,
"And posts, like the commandment of a king,
"Sans check to good and bad: but when the *planets*
"In evil mixture to disorder wander,
"What plagues and what portents! what mutiny!
"What raging of the sea! shaking of earth!
"Commotion in the winds! frights, changes, horrors,
"Divert and crack, rend and deracinate
"The unity and married calm of states
"Quite from their fixure!"

For the use of the word

SPHERE OR ORB

in its Ptolemaic sense frequent examples could be adduced. It will suffice to cite the following:

the sun, the moon, and the stars: as if we were villains by necessity; fools by heavenly compulsion; knaves, thieves, and treachers, by spherical predominance; drunkards, liars, and adulterers, by an enforced obedience of planetary influence; and all that we are evil in, by a divine thrusting on: an admirable evasion of whoremaster man, to lay his goatish disposition to the charge of a star! My father compounded with my mother under the dragon's tail; and my nativity was under Ursa major; so that it follows, I am rough and lecherous. Tut, I should have been that I am, had the maidenliest star in the firmament twinkled on my bastardizing." (Lear, I, II, 128ff.)

[1] heavens=spheres.

[2] this centre=the Earth.

[3] Cp. Batman, New Sh. Soc., ut sup., p. 443: "which Sunne is placed among "the seaven great Starres, called the seaven Planets ..... The Sunne is the fourth "in place, as it were a King in the middest of his throne." For a similar thought compare Dubartas, translated by Sylvester, 1621, p. 84.

"I do wander everywhere, Swifter than the moon's sphere"
(Mids. N. Dr., II, I, 7.)

Certain stars shot madly from their spheres.
(Mids., II, I, 153.)

You stars that move in your right spheres.
(John, V, VII, 74.)

Two stars keep not their motion in one sphere.
(1. Henry IV., V, IV, 65.)

unsphere the stars.
(Wint. Tale, I, II, 48.)

fly . . . . like a star disorbed.
(Troil., II, II, 46.)

Make thy two eyes, like stars, start from their spheres.
(Haml., I, V, 17.)

My good stars . . . . . have empty left their orbs.
(Antony, III, XIII, 146.)

etc. etc.

THE HARMONY OF THE SPHERES.

According to an old theory, first advanced by the Pythagorean school, the Spheres (the seven planets corresponding to the seven notes of the heptachord) moving within one another produced music, 'the harmony of the spheres', though not audible to mortal ears. This belief was taken over by Plato and remained familiar to the medieval mind.[1] It is illustrated by the following passages in Shakespeare's works:—

There's not the smallest orb which thou behold'st
But in his motion like an angel sings,[2]
Still quiring to the young-eyed cherubims;
Such harmony is in immortal souls;
But whilst this muddy vesture of decay
Doth grossly clothe it in, we cannot hear it.
(Merch. of Ven., V, I, 60 ff.)

His voice was propertied As *all the tuned spheres*.
(Ant. & Cleop., V, II, 84.)

The music of the spheres.
(Twelfth Night, III, I, 121, and Pericles V, I, 230.)

Discord in the spheres.
(As You Like It, II, VII, 6.)

[1] Cp., too, Batman, New Sh. Soc., Trans., 1877—"79, pp. 438, 442, 449.

[2] Compare Plato, De Republica (10th book) where he speaks of the harmony of the Spheres, and represents a Siren sitting on each of the eight orbs, and singing to each in its proper tone. From all the eight notes there results a single harmony. (See Furness, VII, pp. 249—250, and Cp. Furnivall's remarks in the New Sh. Soc., Trans., 1877—"79, p. 450.)

THE EIGHT SPHERES.

Each of the seven Planets is mentioned by Shakespeare:

1. THE PLANET MOON:

> I am marble-constant; now the fleeting *moon*
> *No planet* is of mine.
> (Ant. & Cleop., V, II, 240.)
>
> You would lift the moon out of her sphere.
> (Temp., II, I, 183.)
>
> Cp. also Mids. N. Dr., II, I, 7, etc.

The belief, that *insanity* is subject to the changes of the moon, is still shared by many people of our days,—the only remnant, it appears, of the old astrological faith. Hence words like: lunatic, lunacy, lune—also used by Shakespeare.

The moon was supposed to have influence on the *growth of plants.* Compare:

> As true as steel, as plantage to the moon.
> (Troil., III, II, 184.)

I need scarcely add that the dependance of the *tides* upon the influence of the moon was known to Shakespeare:

> you may as well
> Forbid the sea for to obey the moon.
> (Wint. Tale, I, II, 426.)
>
> . . . that ebb and flow by the moon.
> (Lear, V, III, 19.)
>
> the moist star[1]
> Upon whose influence Neptune's empire stands.
> (Hamlet, I, I, 118.)

Compare also:—

> watery moon
> (Mids., II, I, 162; and Rich. III., Act II, II, 69);
> the watery star
> (Wint. Tale, I, II, 1).
> etc.

[1] 'watery' and 'moist', with reference to the influence on tides. The moon is called 'moist' already in Lydgate's 'Story of Thebes', Prologue. The following passage from Bacon (ed. Ellis and Spedding, vol. V, p. 550) might possibly point to another interpretation: "[this theory of mine] denies that *the moon is either a* "*watery or a dense or a solid body;* affirming that it is of a flamy nature." [The italics are in the original text.]

2. THE PLANET MERCURY.

My father named me Autolycus; who being, as I am, littered under Mercury, was likewise a snapper-up of unconsidered trifles.

(Wint. Tale, IV, III, 25.)

3. THE PLANET VENUS.

Yet you, the murderer, look as bright, as clear,
As yonder Venus in her glimmering sphere.

(Mids. N. Dr., III, II, 60.)

Compare, too, Mids. N. Dr., III, II, 107; 2. Henry IV., Act II, IV, 286; 1. Henry VI., I, II, 144; Titus Andron., II, III, 30.

4. THE PLANET SUN.

The glorious planet Sol.

(Troilus, I, III. 89.)

O sun, Burn the great sphere thou movest in!

(Ant. and Cleop., IV, XIII.)

"The Sunne is the eye of the worlde", says Batman, *ut. sup.*, p. 443. 'Oculus mundi' is the Latin for it. Compare Shakespeare's "Heaven's eye". (Com. of Err., II, I, 16; Titus Andr., II, I, 130; IV, II, 59; comp. Rich. II., Act III, II, 37.)

5. THE PLANET MARS,

whose motions formed a puzzle to the astronomers, until they were correctly explained by Kepler. Nash, in "Have with you to Saffron-walden", 1596, (Huth Libr. ed., Nash, vol. III, p. 28) says, e. g., "you "are as ignorant in the true movings of my Muse as the Astronomers "are in the true movings of Mars, which to this day they could never "attaine too."

Compare:

Mars his true moving, even as in the heavens
So in the earth, to this day is not known:
Late did he shine upon the English side;
Now we are victors. (1. Henry VI., I, II, 1.)

Again, All's Well, I, I, 210,

Helena. you [Parolles] must needs be born under Mars.
Parol. When he was predominant.
Hel. When he was retrograde, I think, rather.

6. THE PLANET JUPITER.

Jupiter: . . . our Jovial star reign'd at his birth

(Cymb. V, IV, 105).

Again—

Your letter is with Jupiter

(Titus, IV, III, 64.)

Here Jupiter *may* refer to the planet.—

7. THE PLANET SATURN.

Falst. Kiss me, Doll.
Prince. Saturn and Venus this year in conjunction!

On this passage Johnson comments: This was, indeed, a prodigy. The astrologers, says Ficinus, remark, that Saturn and Venus are never conjoined.

Saturn caused hate, melancholy, and moroseness, compare the word 'saturnine'.

Madam, though Venus govern your desires, Saturn is dominator over mine, *etc.*

so says Aaron, in Tit. Andr., II, III, 31.

Similarly, Conrade the follower of Don John, in Much Ado Abt. Nothing, I, III, 12, is said to have been "born under Saturn".

8. THE EIGHTH SPHERE,

called the *Firmament*,[1] contained the *Fixed Stars* with their Constellations.

Of THE ZODIAC (in this Sphere), referred to twice—

The sun . . . gallops the zodiac in his glistering coach

(Titus, II, I, 7);

nineteen zodiacs [=years] have gone round

(Meas. f. Meas. I, II, 172)—

several constellations are mentioned:—

I) ARIES=THE RAM;

II) TAURUS=THE BULL.

Compare Tit. Andr., IV, III, 64–72, where the 'distract' Titus bids his companions shoot arrows towards heaven:

Tit. Good boy, in *Virgo's* lap; give it Pallas.[2]
Marc. My lord, I aim a mile beyond the moon;
Your letter is with Jupiter by this.
Tit. Ha, ha!

---

[1] "This brave o'erhanging firmanent, this majestical roof fretted with golden "fire", Hamlet, II, II, 312.

[2] Schmidt, in his lexicon, erroneously explains Pallas to be a planet. N. B. the planetoid Pallas was only discovered in 1802!

Publius, Publius, what hast thou done?
See, see, thou hast shot off one of *Taurus'* horns.
Marc. This was the sport, my lord: when Publius shot,
The Bull, being gall'd, gave *Aries* such a knock
That down fell both *the Ram's* horns in the court.

Taurus is also mentioned in Twelfth Night, I, III, 146:

born under Taurus.

III) CANCER.

Add more coals to Cancer when he burns
With entertaining great Hyperion.
(Troilus and Cress., II, III, 206.)

The reference is, of course, to the summer solstice.

IV) VIRGO (Tit. Andr., ut sup., v. 64).

For ASTROLOGICAL PURPOSES the twelve zodiacal signs used to be divided into four TRIGONS,—a Trigon or Triplicity being a combination of three signs in the form of a triangle, each 120° apart. Thus Aries, Leo, and Sagittarius form the first triplicity; Taurus, Virgo, Capricorn the second, and so forth.[1] Each Trigon was denominated from the connatural element: so there are three fiery, three watery, three airy, and three earthly signs:

Fiery: Aries, Leo, Sagittarius.
Airy: Gemini, Libra, Aquarius.
Watery: Cancer, Scorpio, Pisces.
Earthly: Taurus, Virgo, Capricornus.

Shakespeare calls the red-nosed Bardolph a fiery Trigon, in 2. Henry IV., II, IV, 285:—

Fal. Kiss me, Doll.
Prince. Saturn and Venus this year in conjunction! *what says the almanac* to that?
Poins. And, look, whether *the fiery Trigon,* his man, be not lisping to his master's old tables, his note-book, his counsel-keeper.[2]

Each of the zodiacal signs was supposed to GOVERN A CERTAIN PART OF THE HUMAN BODY, or to put it in Chaucer's language: "everich of thise 12 Signes hath respecte to a certain parcelle of the

[1] See Skeat's Introduction to Chaucer's Astrolabe (pp. LXVI—LXVII, Chauc. Soc. XXIX), & Nares's Glossary.

[2] "Aries, Leo, and Sagittarius are hot, drie, and bitter, Cholerike, and are "governyng hot and drie thinges, and *this is called the fierie triplicitie.*" (Bullein's Dialogue, 1578, Early Engl. Text Soc., No. LII, p. 32.)

"body of a man and hath it in governance: as aries hath thin heved, "& taurus thy nekke & thy throte, gemini thyn armholes & thin "armes, & so forth".[1] Let me also quote from Nash's Prognostication: "Taurus, which governes the neck and throat".[2]

Compare Twelfth Night, I, III, 146ff.:

Sir Toby. were we not born under Taurus?
Sir Andr. Taurus! that's sides and heart.
Sir Toby. No, sir; it is legs and thighs. Let me see thee caper, etc.

Sir Toby's blunder: "it is legs and thighs" is, of course, intentional, as he wishes to make Sir Andrew believe he was born under a dancing star; while Sir Andrew's ignorance seems to be genuine. The mention of Taurus at all seems to be the product of mere caprice, unless, indeed, we suppose Sir Toby to have had in mind another thought, which, however, Sir Andrew failed to understand. For Taurus was imagined by astrologers to be the 'mansion' of Venus and was moreover a 'feminine' sign. Compare Chaucer, Wif of Bath, Prol., v. 613:—

Myn ascendent was Taur.

So much for the zodiac.

Of the same *sphaera stellarum fixarum* three prominent

CONSTELLATIONS

near the North Pole are referred to by the poet: THE GREAT BEAR (or Charles's Wain), THE LESSER BEAR, and the DRAGON.

Charles' Wain is over the new chimney.
(1. Henry IV., Act II. I, 2.)

My nativity was under *Ursa major.*
(Lear, I, II, 141.)

The wind-shaked surge, with high and monstrous mane,
Seems to cast water on the *burning bear,*
And quench the *guards of the ever fixed pole.*
(Othello, II, I, 13.)

---

1 Chaucer's Astrolabe, Part I, § 21.—Compare Plate VII, in Skeat's ed. (Chauc. Soc.), where a human body with the constellations in the different parts and organs they govern is figured. Furness (New Var. Ed., XIII, p. 53) points out that some *almanacs* even of the present day have preserved figures of a similar kind. Compare with this interesting fact Prince Henry's question "what says the "almanac to that", 2. Henry IV., Act II, IV, 287. See the whole passage quoted above.—An astrological Calendar or "*Ephimerides*" is referred to by Nash (ut infra, p. 143). See also "The Kalender of Shepherdes", reprinted by Oskar Sommer, 1892.

2 "A Wonderfull . . . Prognostication", 1591.—Huth Libr. ed, vol. II, p. 155.

The burning bear probably=the Great Bear; pole=the polar star, the guards of which are the two stars known astronomically as Beta and Gamma in the constellation of the Lesser Bear ('on the shoulder 'and foreleg, as usually depicted, or sometimes on the ear and shoulder'). Compare Blundevile, in His Exercises (4th ed., 1613), p. 716: "Of which seven starres *in the little Beare*, two starres called the "*guards of the North starre*, are to the Mariners most familiar."[1]

THE DRAGON:

> My father compounded with my mother under the *dragon's tail;* and my nativity was under Ursa major; so that it follows, I am rough and lecherous.
>
> (Lear, I, ii, 140.)

"THE SEVEN STARS",[2] mentioned in Lear, I, v. 38; 1. Henry IV., Act I, ii, 16; and in 2. Henry IV., Act II, iv, 201, are the Pleiads, or perhaps the Great Bear.

Having noted the constellations, it remains to name two FIXED STARS referred to by the poet:—

i) THE POLE-STAR.

> I am constant as the northern star,
> Of whose true-fix'd and resting quality
> There is no fellow in the firmament.
>
> (Julius Caesar, III, i, 60.)

the north star (Much Ado, V, i, 258);
the lode-star (Mids. N. Dr., I, i, 183; Lucrece, 179);
the pole [=the pole-star] (Hamlet, I, i, 36);
the ever fixed pole (Othello, *ut sup.*).

ii) To the influence of the DOG-STAR=Sirius (or Procyon) were attributed the 'dog-days':

> Twenty of the dog-days now reign in's nose.
>
> (Henry VIII., Act V, iv, 43.)

In Love. Lab. L.; Rich. II.; and Henry VIII., 'fixed stars' in general are referred to.

As to the

SUBSTANCE OF THE STARS,

compare what Bacon says:[3] "Another question is, *are the stars true "fires?*"—"Aristotle . . . had determined the stars to be real fires."

[1] For fuller information, see Blundevile, pp. 716—717.
[2] Cp. New Shaksp. Soc. Transactions, 1877—79, p. 448 note.
[3] Bacon, ed. Ellis and Spedding, vol. V., pp. 538—9.

Compare:—

Doubt thou *the stars are fire;*
Doubt thou the sun doth move;
Doubt truth to be a liar;
But never doubt I love.

(Hamlet, II, II, 116.)

SPHERE OF FIRE.

Round the Atmosphere environing the Earth (and Sea) there was imagined to be a Region of Fire, as I have said above. "Fire", to quote from Blundevile, "being placed next the sphere of the Moon "under which it is turned about like a celestial Sphere."[1]

Compare 'Hamlet', V, I, 305:—

let them throw
Millions of acres on us, till our ground,
Singeing his pate against the *burning zone,*
Make Ossa like a wart!

By the action of this Fire, in some manner or other, "fierie "impressions, as lightnings, fire drakes, blazing starres and such like"[2] were thought to be bred in the highest region of the air.

A METEOR

was supposed to be a kind of vapour drawn up by the sun and acted upon by the hot influences in the upper air. Hence, Exhalation=Meteor.

It is some meteor that the sun exhales.

(Romeo, III, v, 13.)

an exhaled meteor.

(1. Henry IV., Act V, I, 19.)

I shall fall
Like a bright exhalation in the evening,
And no man see me more.

(Henry VIII., Act III, II, 226.)

---

[1] Compare Sylvester's Translation of Du Bartas, 1621, p. 25:
"But, lest the Fire (which all the rest imbraces)
"Being too neer, should burn the Earth to ashes;
"As chosen Umpires, the great All-Creator
"Between these Foes placed the Aire and Water."

[2] Blundevile, *ut sup.*, p. 377. As for comets, I think Tycho de Brahe was the first, who demonstrated them to be located far beyond the orbit of the moon. Bacon distinguishes between sublunar and higher comets. But cp. Batman, ut sup., p. 446.

Meteors are called, too, by the name of *shooting stars*, *falling stars*, or *fire-drakes*.

Compare:

> I see thy glory like a shooting star
> Fall to the base earth from the firmament.[1]
> (Rich. II., Act II, iv, 20.)
>
> How a bright star shooteth from the sky.
> (Venus and Ad., 815.)
>
> These are stars indeed;
> And sometimes falling ones.
> (Henry VIII., Act IV, i, 55.)
>
> that fire-drake.
> (Henry VIII., Act V, iv, 45.)

COMETS

or blazing stars.

Compare:

> Comets, importing change of times and states,
> Brandish your crystal tresses in the sky.
> (1. Henry VI., Act. I, i, 2.)
>
> but one every blazing star.
> (All's Well, I, iii, 91.)

Examples might be multiplied.

Perhaps I ought to make mention, too, of

ST. ELMO'S FIRE

here, to which there is an apparent allusion in The Tempest, I, ii, 196 seq. Whatever Shakespeare thought about its cause, it was explained to be "an exhalation of moyst vapours, that are ingendred "by foule and tempestuous weather".[2] Linschoten says in his 'Discourse of Voyages' (Hakluyt Soc., *71*, p. 238):—

> [Somewhere off the coast of Natal] wee lay driving for the space of two dayes and two nights together, with a continuall storme and fowle weather with rayne. The same night we saw uppon the main yarde, and in many other places a certaine signe, [called] ..... St. Elmo ..... *This constellation (as Astronomers doe write)* is ingendred of great moisture and vapors, and sheweth like a candle that burneth dimly, and skippeth from one place to another, never lying still.

[1] Firmament may be equivalent to sky, heaven. It need not necessarily mean the eighth sphere.

[2] Purchas his Pilgrimes; see Clarend. Press ed. of The Tempest, p. 86.

SECTION 3.—SHAKESPEARE'S ENVIRONMENTS REFLECTED IN HIS WORKS.

Wer den Dichter will verstehn
Muss in Dichters Lande gehn.
Goethe.

As for place-names mentioned in Shakespeare's text, it is significant that no country scores higher than England, the home of the great national poet. Of English places-names (counting each once) we find about 210 in Shakespeare's plays. About half of these are London and Warwickshire names or names of places very near to either. Of these I wish to give a full list, as a knowledge of Shakespeare's immediate environments reflected in his works will help us to get into closer touch with him. And we want to get as close to him as possible. We want to know where *his* feet walked and what *his* eyes saw.

To

LONDON,

in which so many scenes of the historical plays are laid there will be found a surprizingly large number of local allusions in Shakespeare's works. An entertaining work on "Shakespeare's London" has been written by Mr. T. F. Ordish, which book (though neither exhaustive nor above criticism) used with a constant reference to Ryther's map of London, *1604*, (reproduced in Loftie's History of London) will serve as an excellent introduction to a concrete knowledge of Shakespeare's London.[1]

THE CITY AND ITS IMMEDIATE VICINITY.

N. B. The *asterisk* denotes that Shakespeare did *not* find the name or a clear suggestion for it in the known source or sources of the play where it occurs.

Baynard's Castle (Rich. III); to the **bear-gardens* of London there are allusions in Shakespeare's text, especially in Merry Wives, I, I, and 2. Henry VI., V, I, 144 *seq.*; **Bedlam* (2. Henry VI., Lear); **St. Bennet* (Twelfth Night); *Blackfriars* (Henry VIII.); *Blackheath*

[1] The final authority on Shakespeare's London is John Stow's "Survey of London" which sadly needs editing *with index* and *illustrative plans.* The student should also refer to Halliwell-Phillipps's Handbook-Index, p. 547. and to Wheatley, London Past and Present.

(Henry V.); the Boar's-Head Tavern[1], stated to be the scene of the merry pranks of the madcap Prince and Falstaff in 1. and 2. Henry IV. is nowhere mentioned in Shakespeare's original text. "*Eastscheap*" (1. and 2. Henry IV.), where this inn was, "and Fish "street Hill abounded with taverns in Shakespeare's time". Ordish, *ut sup.*, p. 35; **Bucklersbury* (Merry Wives, III, III, 79), a street of London, chiefly inhabited by druggists; **Charing-cross* (1. Henry IV., II, I, 27); *Cheapside* (2. Henry VI.); "*the *Countergate* (Merry Wives, III, III, 85).—The reference is to one of the prisons called the Counter. There were two of that name in the City and one in Southwark. There may be an allusion, too, to the Counter in Errors, IV, II, 39. *Crosby Place* (Rich. III.); *Ely House* (Rich. II.)—the Bishop of Ely's palace in Holborn, the site of which is still marked by 'Ely Place'. In Rich. III., Act III, IV, 33, Richard addressing 'my Lord of Ely' says: 'when I was last in *Holborn*, I saw good strawberries in your garden 'there' (see Clar. Press ed.); **Finsbury* (1. Henry IV.); 'up *Fish Street*! 'down *Saint Magnus'* Corner' (2. Henry VI.); *the Fleet* (2. Henry IV.) a prison in London; "**the windmill in St. George's Field*", to the South of London (2. Henry IV.); *Guildhall* (Rich. III.); Shakespeare has several allusions to the *Inns of Court* (the four sets of buildings in London, the Inner Temple, the Middle Temple, Lincoln's Inn, and Gray's Inn): "He is at Oxford still, is he not? .... A' must, then, "to the **inns o' court* shortly. I was once of **Clement's Inn.*" (2. Henry IV., III, II, 11—15). **Clement's Inn* (mentioned several times, (2. Henry IV.), an inn of chancery, was subordinate to the Inner Temple. "All the **inns o' court*" (2. Henry IV., III, II, 25); "*the inns of court*" (2. Henry VI., IV, VII, 2); "We sent unto **the Temple*, unto his Chamber" (2. Henry VI., II, V, 19); "in *the *temple-hall*"[2] (1. Henry IV., III, III, 223); "Within the **Temple-hall* we were "too loud; The **garden here* is more convenient" (1. Henry VI., II, IV, 3—4); "in the **Temple-garden*" (1. Henry VI., II, IV, 125); "behind **Gray's Inn*" (2. Henry IV., III, II, 36).—**Limehouse* (Henry VIII.). *London* is, of course, frequently mentioned. *London bridge* (1. and 2. Henry VI.; Rich. III., III, II, 72); "*London gates*" (at either end

[1] As to the Boar's-Head Tavern cp. Clarend. Press ed. of 1. Henry IV., p. 129, Ordish, Shakespeare's London, p. 35, Halliwell-Phillipps, Outlines of the Life of Shakespeare (index).

[2] Remember that in the Gothic 'Hall' of the Middle Temple, built in 1572, Shakespeare's Twelfth Night was acted on Febr. 2., 1601—2. In the hall of Gray's Inn the Comedy of Errors was played on Decemb. 28., 1594.

of the bridge) in 1. Henry VI; "*London road*" (1. Henry IV.), **London streets* (Rich. II., 2. Henry VI.); *London-stone* (2. Henry VI.); *Lud's town*, ancient name of London (Cymb.); **Lumbert street* (Ff. Lombard.—2. Henry IV., II, I, 31); "*A *Marshalsea*" (Henry VIII.), 'The Marshalsea formerly stood near St. George's Church, Southwark'. 'It now no longer exists', Clar. Pr. ed. of Henry VIII. **Mile-end (Green)* (All's Well, 2. Henry IV.). 'The usual exercise ground of 'the London trainbands'; **Moorfields* (Henry VIII.), 'a place of resort 'where the trainbands of the city used to be exercised'; **Moor-ditch* (1. Henry IV.) in Moorfields; **Newgate* (1. Henry IV.) name of a prison in London; *"*this wooden O*" (Henry V.) = the Globe Theatre; **Paris-garden* (Henry VIII.); St. Paul's Cathedral[1]=**Paul's* (1. Henry IV., II, IV, 576; 2. Henry IV., I, II, 58; Henry VIII., V, IV, 16), *Paul's* (Rich. III., I, II, 30; III, VI, 3); **Pie-corner* (2. Henry IV., II, I, 28); **Pickt-hatch* (Merry Wives, II, II, 19); *The Rose, within the parish Saint Lawrence Poultney* (Henry VIII., I, II, 152); *The Savoy* (2. Henry VI.); *Smithfield* (2. Henry VI., II, III, 7); **Smithfield* (2. Henry IV., I, II, 56 & 59; 2. Henry VI., IV, V, 10 and VI, 14); "in the **south* "*suburbs, at the Elephant*, Is best to lodge"—sounds like some local allusion (Twelfth Night, III, III, 39); *Southwark* (2. Henry VI.); **Star-chamber* (Merry Wives), at Westminster; **the Strand* (Henry VIII.); the *Thames* (2. Henry VI., IV, VIII, 3), **Thames* (Merry Wives; Henry V., IV, I, 120). *The Tower* is very frequently mentioned (it is called "Julius Caesar's ill-erected tower", in Rich. II., V, I, 2); **Tower-hill* (Henry VIII.); **Turnbull street* (2. Henry IV.)—now, and indeed originally, Turnmillstreet, near Clerkenwell (it was the whores' quarter).—"Thou makest the triumviry, the corner-cap of society, The "shape of Love's **Tyburn that hangs up* simplicity". (Love's Lab. L., IV, III, 53); *Westminster* (Rich. II.; 2. Henry IV.; 2. Henry VI.; Rich. III.; Henry VIII.); **the cathedral church of Westminster* (2. Henry VI., I, II, 37); at *Westminster Hall* Richard II. is deposed (Rich. II., Act IV.); ["*westward-ho*" (Twelfth Night, III, I, 146)—an exclamation often heard on the Thames]; "You must no more call it *York-place*, "that's past ..... 'Tis now the king's, and call'd *Whitehall*" (Henry VIII., IV, I, 95); *White-Friars* (Rich. III.); the *White Hart in Southwark*" (2. Henry VI., IV, VIII, 25).

[1] Shakespeare saw this immense church without its original steeple, which had been consumed by lightning in *1561*.

PLACES WITHIN 25 MILES FROM THE CITY:—

St. Alban's (2. and 3. Henry VI.; Rich. III.), **St. Alban's* (1. Henry IV.), "as common as the way between *Saint Alban's and London" (i. e. the Watling Street) (2. Henry IV., II, II, 185); *Asher House* (Henry VIII.)—Asher was the old form of Esher, near Hampton Court, (Clar. Pr. Ed.); *Barnet* (3. Henry VI.); **Brentford* (Merry Wives)—the Elizabethans spelt Brainford or Braynford; *Chertsey* (Rich. III.); *Eltham* (1. Henry VI.); **Colebrook (v. post)*; *Gadshill* (between Gravesend and *Rochester*) (Henry IV.); *Greenwich* (Henry VIII.); *Hatfield* (?), William of (2. Henry VI.); *Langley*, Edmund of (2. Henry VI.). In the 'Merry Wives', the scene of which is * *Windsor*, we have the following names (not adopted from any known source): the *Thames*, the *Datchet-lane*, *Datchet-mead*, *Frogmore*, the *Castle*, the *Park* with *Herne's oak*, *Eton* just across the Thames. *Colebrook* [=Colnbrook] *Maidenhead* and *Reading* are on the way from London to Oxford round Windsor. — *Windsor* is mentioned, too, in (1. Henry IV.; 1. and 2. Henry VI.). At *Windsor Castle* the 3rd, and probably the 6th, scene of Act 5 in 'Rich. II.' is laid. In Henry the Fourth, First Part, I, III, the scene is by the editors supposed to be London, the palace. This is certainly an error. The scene is *Windsor*, cp. Holinshed, and Act I, I, 104.—* *Windsor* (2. Henry IV.). In Twelfth Night "the bed "of * *Ware*" is mentioned. This huge bed still existing is figured in Knight's edition of Shakespeare and in Chambers's Book of Days. In conclusion, I ought to add that the shire-names of Buckingham, Essex, Kent, and Surrey occur in Shakespeare's text. [Perhaps I ought to mention, too, as a locality which Shakespeare seems to have known from autopsy: *Dover* with its castle, the "chalky cliffs" (2. Henry VI., III, II, 101; Errors, III, II, 129), and the cliff, now called Shakespeare's Cliff, "whose high and bending head Looks fearfully in the confined "deep" (Lear).]

WARWICKSHIRE NAMES.

**Barson* (2. Henry IV., V, III, 94) prob.=Barston; **Burton-heath* (The Shrew, Induction) supposed to be Barton-on-the-Heath; *Coventry* (Rich. II., 2. Henry IV., 3. Henry VI.); **Coventry* (1. Henry IV., IV, II, 1 & 42); **Dunsmore* and **Southam* (3. Henry VI.) are not in Holinshed nor in Hall; *Killingworth* (=Kenilworth) (2. Henry VI.); **Sutton Co'fil'* (1. Henry IV.)=Sutton-Coldfield (see Clar. Pr. ed. of 1. Henry IV.); *Tamworth* (Rich. III.); *Warwick*, Earl of, (2. Henry IV., 2. and 3. Henry VI., Rich. III.); *Warwickshire* (2. and 3. Henry VI.), * *War-*

wickshire (1. Henry IV.); * *Wincot* (The Shrew) &* *Woncot* (2. Henry IV.) are usual corruptions of Wilmcote (cp. Knight's edition, vol. I, p. 282).

Near the borders of Warwickshire are the following place-names: **Banbury* (Merry Wives) through which Shakespeare must have passed on his shortest road to London from Stratford; *Bosworth Field* (Rich.III.); **Cotswold* or **Cotsall* (Rich. II., 2. Henry IV., Merry Wives); **Daventry* (1. Henry IV.)=*Daintry* (3. Henry VI.); **Hinckley* (2. Henry IV.), partly in Warwickshire.

Of the rest of the names from English Geography I have made a list for myself. But it is unnecessary to publish it. I count *112* [—122] names exclusive of those given in the above list, which latter amount to *98* [—102]. So that the sum total is *210* [—224].

Of *Wales* I counted 7 names.

Of *Scotland*, 12 (but more, if we add all the names of the noblemen like Angus, Caithness, etc.).

Ireland: 2 names.

SECTION 4.—THE CONTINENT OF EUROPE.

In Elizabethan days

TRAVEL

was the fashion. It was the time when young men went

> Some to the wars, to try their fortune there;
> Some to discover islands far away.
>
> (The Two Gentlemen, I, III, 8—9.)

And

> To see the wonders of the world abroad.
>
> (*ut sup.*, I, I, 6.)

The Elizabethans evidently concurred with Antonio, that one

> cannot be a perfect man
> Not being tried and tutor'd in the world.
>
> (*ut sup.*, I, III, 20.)

On the other hand the poet ridicules the craze for foreign travels (cp. As You Like It, IV, I, 22—40) and the mannerisms of a traveller, 'the picked [=affected] man of countries', who had made the grand tour and could talk

> . . . . of the Alps and Apennines,
> The Pyrenean and the river Po.
>
> (John, I, I, 189 *seq.*; cf., too, All's Well, II, v, 30.)

Of course, Shakespeare must have made some tours in England. Reminiscences of these will be found in his Sonnets. It has been supposed, too, that he visited Scotland, France, Holland, Denmark, Germany. But this is all conjecture. Another question, which has been seriously debated especially by German Shakespearean scholars, is whether the poet was in Italy. To my mind the positive conclusions arrived at by some are not borne out by the existing premises. The strongest reasons, it is said, for supposing Shakespeare to have travelled to Italy are to be found in 'The Merchant of Venice' and 'Othello'. Now, the former play is generally admitted to be founded on an old drama, and 'Othello' I strongly presume is likewise recast from an earlier play. The strongest arguments would therefore turn out to be baseless fabrics. I myself have no doubts that Shakespeare never was in Italy. On the whole, it may be fairly said that the Italy of Shakespeare's plays is Italy reflected in Elizabethan London.

There are two famous discrepancies in Shakespeare's Italian Geography, that require to be noticed: the *tide* at Verona, whence the characters go to Milan *by ship*! in 'The Two Gent. of Verona', and the sea-voyage from Sicily to the *shores of Bohemia*! in 'The Wint. Tale'. In the former case Shakespeare had no serious thought of locality, London being the real background of the play. In the latter case Shakespeare simply adopted a harmless, it may be consciously made, error from Greene's 'Pandosto', the source of 'Winter's Tale'.

ITALY.

As the Italian Geography in Shakespeare's plays makes considerable claims on our attention and interest, I have given in full all the Italian names of places of modern Geography:—*Apennines* (John); *Bergamo* (The Shrew); *Calaber*, Duke of (2. Henry VI.)=Calabria, (Holinshed has *Calabre*); *Etna* (Merry Wives); *Ferrara* (Henry VIII.); *Florence*, *Florentine* (The Shrew, All's Well, Othello); *Genoa* (Merchant, The Shrew); *Italy*, *Italian*, etc. *(passim)*; "fruitful *Lombardy*, The "pleasant garden of great Italy" (The Shrew, I, I, 3); *Mantua* (Two Gentlemen, The Shrew, Romeo, Love's Lab. L.); *Messina* (Much Ado), in Sicily; *Milan* (Tempest, Two Gentlemen, Much Ado, John); *Naples*, *Neapolitan* (Tempest, etc.); *Padua* (Much Ado, Merchant, The Shrew) "Padua, nursery of arts" (The Shrew, I, I, 2); *Pisa* (The Shrew); *Po* (John); *Rome, Roman (passim)*; *Senoys* (All's Well)=*Siennese; Sicil, Sicilia, Sicily* (2. Henry IV., Wint. Tale, etc.); *Venice, Venetian* (Much Ado, Love's Lab. L., etc.), *the Rialto* in Venice—the market-place, not the bridge—(Merchant); *Verona, Veronessa* (Two Gentlemen, The Shrew, Romeo, Othello).

On the opposite side of the Adriatic is *Illyria* (Twelfth Night). *Messaline* mentioned in Twelfth Night is a name of place unknown in Geography.[1]

IN CONCLUSION.

In the dramas dealing with English History the scene is of course on English or French soil. The Geography is therefore the same as in the Chronicles. In Macbeth the scene is Scotland and England. In King Lear we are in old Britain, which, too, is partly the locale of Cymbeline. The dramas based on Plutarch, as well as Titus Andronicus, Com. of Errors, Troilus, and Pericles take us to classic ground. In seven of the rest of the dramas pronouncedly romantic the scene is Italy, in two (Love's Lab. L., As You Like It) France, in Twelfth Night Illyria, in Hamlet Denmark, in Meas. for Measure the German Empire, in The Merry Wives England. The Prospero-Island of the Tempest is some undefined isle in the Mediterranean. The Winter's Tale takes us both to Sicily and Bohemia; Cymbeline to Britain and Italy; Othello to Italy and Cyprus; All's Well to France and Italy.

[1] Of the rest of the countries of Europe I offer only the arithmetical totals of geographical names here:—*Islands of the Mediterranean:* 5 names;—*France:* circa 50;—*German Empire*: 13;—*The Netherlands*: 8;—*Denmark, Norway, Iceland:* 4;—*Spain, Portugal:* 5;—*Turkey, Greece:* 7;—*Switzerland, Hungary, Poland, Russia, Lapland*: 8. So far I have treated modern Geography. Of Antique Geography (both European and non-European) I count 117 names.

For Information on Europe in Shakespeare's time, see Fynes Moryson's "Itinerary", 1617. Extracts from the Fourth Part (never printed before) of this Itinerary were published by Mr. Hughes, under the title of "Shakespeare's Europe". 1903.

APPENDIX TO CHAPTER 7.

1.—A NOTE ON MAPS AND GLOBES.

"THE NEW MAP."

Twelfth Night (Act III, sc. II, 84—86) contains a distinct allusion to a certain contemporary map:—

Malvolio, we are told, "*does smile his face into more lines than is in* "*the new map with the augmentation of the Indies*".

To begin with, what is meant by 'more lines', 'augmentation', and 'the Indies'?

The *many lines* undoubtedly point to a map crossed with loxodromes or rhumb-lines, characteristic of so many early maps, especially of charts. The map, therefore, must have been on the cylindrical projection [i. e., a map on Mercator's projection, or a map on the 'equidistant cylindrical projection', or a compass map, which would belong to this class too], which alone admits of the straight rhumbs.

The words *'augmentation'* and *'the Indies'* are best explained by the following part of a title of a treatise by Thomas Blundevile: "A plaine and full description of Petrus Plancius his universall Map, serving both for Sea and Land, and by him lately put forth in the yeare of our Lord. 1592. *In which Mappe are set downe many more places, aswell of both the Indies* as of Afrique, ... then are to be found eyther in Mercator his Mappe, or in any other moderne Mappe whatsoever" ... M. Blundevill ... 1594 ... London.[1]

[1] Plancius's map, designed to supersede Mercator's famous map of 1569, is unhappily lost. From Blundevile's description and from a notice in the Dutch State papers (see De Jonge, 'Opkomst', I, 167) we know that the map had on it rhumbs. Though it was nine years old in 1601, I at one time thought that it was Shakespeare's map. But I am diffident about this now. — As a pendant to the above title, I quote another passage from Blundevile (Exercises, 4th ed., p. 514): "the Mappe which covereth M. Molineux his Terrestrial *Globe*, differeth "greatly from Mercator his terrestial *(sic)* Globe, by reason that there are found "out *divers new places* aswell towards the Northpole, *as in the East and West* "*Indies*, which were unknowne to Mercator".

'Indies' I thus take to mean *both* the Indies, that is to say *America* generally and the large portion of *South-Eastern Asia*. Now, both of these parts of the world had made their appearance in the maps of the world long before the end of the sixteenth century. 'Augmentation' cannot therefore mean *addition*, as the New English Dictionary and Schmidt's Shakespeare Lexicon would make out, but must be equivalent to *enlargement, increase*.

Various attempts have been made to identify the Shakespearean map. First, in point of time, comes George Steevens, who supposed that Shakespeare alluded to a map of the Eastern Islands, which was contained in Linschoten's Voyages, an English translation of which appeared in 1598. Steevens's explanation is wholly unsatisfactory. By far more plausible is a theory advanced by James Lenox in 1860, and vigorously defended by C. H. Coote in a paper read before the New Shakspere Society, 1878 (Trans. 1877—9, Pt. I, p. 88), and again in the Introduction to the 59th volume published by the Hakluyt Society. Lenox and Coote identify the 'new map' with the map ("a true hydrographical description of so much of the world as hath beene hetherto discovered") contained in several copies of the second edition of Hakluyt's "Principal Navigations", 1598-1600 (vol. III, 1600), and presumably also published separately.[1] There is reason to suppose that Molyneux, Wright, and Hakluyt were the joint constructors of this important map.

The map is multilineal in the extreme and new. New, not only because made about 1599 and giving geographical information up to the knowledge of the time, but because it is laid down upon an improved Mercator-projection, as Mr. Coote points out. It remains to show whether it has the required 'augmentation of the Indies'. As Mr. Coote's arguments bearing on this question have been repeatedly referred to and cited as authoritative, it will be worth while quoting them here again literally. After having stated that he excludes the West Indies from his considerations and having given, what appears to me, a somewhat forced interpretation of the word 'augmentation', he proceeds thus:[2]

"Now what was the state of things to be seen upon the eastern "portion of our 'new map' at the close of the 16th century, as compared "with all the best general maps of the world that preceded it? A marked "development in the geography of India proper, then known as the land "of the Mogores or Mogol, the island of Ceylon, and the two peninsulas "of Cochin China, and the Corea. For the first time the distant island of "Japan began to assume its modern shape (this last, by the way, is not

---

[1] A reproduction of this map accompanies the 59th volume published by the Hakluyt Society, and is also contained in A. E. Nordenskiöld's Facsimile Atlas, Tab. L. In the New Shakspere Soc., *ut sup.*, p. 88, the East Indian Archipelago is unfortunately excluded from the section (chiefly Asia) there printed.

[2] New Shakspere Soc. *ut sup.*, p. 95.

"to be seen on the map in Linschoten). Turning to the S. E. portion of "the "new map" (unfortunately not shown in the section before you), "there were to be seen traces of the first appearance of the Dutch under "Houtman at Bantam (W. end of Java), synchronizing almost within a "year with that of their fellow-countrymen in Novya Zembla; and which "within 10 years led to their unconscious discovery or rather rediscovery, "of Australia.

"On all the old maps, including the one of Ortelius's inserted in our "old friend Linschoten, was to be seen the huge Terra Australis of the "old geography. This, as Hallam remarked, had been left out upon our "'new map', and in its place was partly to be traced New Holland "[=Australia]. This of course would be suggestive of nothing to the mind "of Shakspere; but what is so remarkable is, that upon our 'new map' "there should have appeared to rise, like a little cloud out of the sea, "like a man's hand, the then unknown continent of Australia.

"It is this appreciation of the marked improvement and development "to be observed in the geography of the eastern portion of our map, to "which I believe Shakspere desired to give expression in his judicious "and happy use of the term 'augmentation'."

These remarks are open to the following criticism. If, as Coote supposes, 'Indies' refer only to the East Indies, he should certainly have taken into consideration not merely maps of the world but also maps of Asia alone. "A marked development in the geography of India proper" and "Ceylon" is non-existent. As early as 1561 Gastaldi had drawn a fairly good map of India in his map of Asia. Compare also the advanced delineations of India in Mercator's Mappemonde (1569), in De Jode's "Asia Partium Orbis Maxima" 1593, and the map of India in Linschoten's Itinerario. Coote's statement, by the way, "India proper, then known as "the land of the Mogores or Mogol" requires to be corrected, as only the northern part of the peninsula was subject to the Mogul. (Compare the Atlases of Metellus, Janson, and Blaeu). The great improvement in the geography of Corea, Cochin China, and Japan on Coote's map is not very noticeable or plain, when we compare maps like Plancius's "Orbis Terrarum Typus" (1594) or (setting aside Corea) De Jode's above mentioned map of Asia, and the map of South-Eastern Asia (with the islands) in Linschoten's Itinerario, 1595—6, (contained also in the Latin Translation).[1]

The same remark applies to the delineation of the Malay Archipelago. With regard to Coote's suggestion as to Australia, I think the tract of land, drawn below Java, is nothing more than a remnant of the old

[1] The Portuguese possessed excellent manuscript maps of the Malay Archipelago at a much earlier date, as appears from Diego Homen's map in Ruge's Zeitalter der Entdeckungen, p. 535. The maps, however, were kept secret and remained unpublished.

"Terra Australis". What Coote means by the "traces of the first appear- "ance of the Dutch under Houtman at Bantam", is difficult to say. Bantam had appeared on maps long before 1600. We have to hear in mind, too, that Coote's map is what it professes to be: a hydrographical map, that is, a chart, on which only coast-names are noted. Coote's map has, indeed, comparatively few names. In the Eastern Archipelago, e. g., (including the Philippines) we find but three and twenty place-names. On Mercator's map of the world (1569) there were twenty-seven on *Sumatra alone.* — Even Plancius's Orbis Terrarum Typus has twenty-four. — But *sapienti sat.* Whatever may be thought of the map in Hakluyt's work, Mr. Coote's arguments for it require revision. None the less, I would assert that of the extant maps the Hakluyt-map has the best claims to be considered as Shakespeare's "new map". Still, the possibility remains, that the poet may have meant another map now lost. *Grau ist alle Theorie,* says Goethe. The above map, however, extremely instructive in many respects, is worthy the attention of every friend of Shakespearean geography.[1]

To

MAPS IN GENERAL

Shakespeare repeatedly refers in his text, e. g., "Peering in maps for ports "and piers and roads" (Merchant of Ven., I, I, 19), "All the quarters that "they know *I' the shipman's card*". (Macb., I, III, 17); "he is the card "or calendar of gentry, for you shall find in him the continent of what "part a gentleman would see". (Hlt., V, II, 117). Lucrece, 1712; and so forth. In Henry V., Act IV, sc. VII, 25, "*maps of the 'orld*" are mentioned. In two plays maps are brought on the stage. In 1. Henry IV., Act III, sc. 1, a

MAP OF ENGLAND

is produced by Glendower ("Come, here's the map") for the purpose of pointing out the threefold division of the realm. Hotspur objects to the winding course of the river Trent:

"Methinks my moiety, north from Burton here,
"In quantity equals not one of yours:
"See how this river comes me cranking in", etc.

---

[1] A well known bibliographer, Henry Stevens, evidently not satisfied with the existing theories proposed a third in his *Catalogue*, 1881, p. 200, where he says: "The writer is half [!] inclined to think that the map 'with the Augmentation "of the Indies', may be the curious little round face-shaped map of the World "in Wytfleit's [*sic*] "Ptolemæum Augmentum" [I suppose, he means 'Descriptionis "Ptolemaicae Augmentum'] first published in 1597, and reissued in Latin and "French in 1598, 1603, 1607, and 1611, it [what does *it* refer to?] being simply "Ptolemy's World *Augmented by* the addition of the East and West Indies. The

Again, in Lear (Act I, sc. I)

A Map of Britain

is produced on the stage for another tripartite division of the kingdom.

Shakespeare may have seen, and made use of, any one of a number of

Maps of England and Great Britain.

The following list which I put forward tentatively, will give some idea of the maps existent between 1564 and 1610:

(1) Angliae ... Nova Descriptio ... 1573, contained in Ortelius's Atlas; (2) Typus Angliae. J. Hondius fecit, 1590; (3) Britanniae Insula, Romae 1589; (4) Britanniae Insula Typus ex conatibus A. Ortelii, 1595; (5) Angliae Scotiae et Hiberniae Descriptio in Ortelius's Atlas; (6) Great Britain and Ireland is also in De Jode's Atlas; (7) 1574—79 Christopher Saxton published maps of the Counties of England and Wales. fol.; (8) Mercator, too, published a good collection of maps of the British Isles prefaced by a dedication to Queen Elizabeth in his Atlas of 1595; (9) An Atlas of Great Britain and Ireland by Camden, London, 1607; (10) John Speed is also to be mentioned as a distinguished cartographer. A collection of his maps appeared in 1611, the "Theatre of the Empire of Great Britain".

Globes.

Shakespeare was familiar with globes of his time, as we learn from

I) The Comedy of Errors, Act III, sc. II, 116:

Dro. S. " . . . she is spherical like a globe; I could find out countries in her."

Ant. S. "In what part of her body stands Ireland?", etc.

II) 2. Henry IV., Act II, IV, 309:

"thou globe of sinful continents".

III) Lucrece, 407:

"Her breasts, like ivory globes circled with blue,
"A pair of maiden worlds unconquered."

The question what globes were known to the poet, shall not detain us long. Up to the date of 1592 Mercator's globes (1541 & later) were in common use in England. In this year, 1592, appeared Molyneux's globes said to be the first constructed in England. These were made both on a larger and a smaller scale. Continental globes by van Langeren (mentioned as a globe-maker as early as 1580) or by his later rival, Jodocus

---

"Indies at that time meaning the West as well as the East Indies."—This suggestion is not to be taken seriously. In the editions I saw I did not find a "face-shaped map of the World". If the map Stevens refers to was round, it could not at least have had rhumbs upon it.

Hondius[1], or by Blaeu would be sold in England, too. Copies of globes by every one of the above globe-makers are still extant.

II.—A NOTE ON THE OLD THEORY OF EARTHQUAKES.

Gabriel Harvey, setting forth the views then widely prevalent as to the cause of earthquakes, says[2]:—

Within the Earth a disturbance is brought about by reason of bad humours encountering the good. "Which conflict indureth so long, and is "fostred with aboundaunce of corrupt putrified Humors, and ylfavoured "grosse infected matter, that it must neede, (as well, or rather as ill, "as in mens and womens bodyes) brust out in the ende into one perillous "disease or other."—"The Materiall Cause of Earthquakes, (as is superficially "touched in the beginning of our speache, and is sufficiently prooved by "Aristotle in ye second Booke of his *Meteors*) is no doubt great aboundance "of wynde, or stoare of grosse and drye vapours, and spirites, fast shut "up, & as a man would saye, emprysoned in the Caves, and Dungeons of "the Earth: which winde, or vapors, seeking to be set at libertie, and to "get them home to their Natural lodgings, in a great fume, violently rush "out, and as it were breake prison, which forcible Eruption, and strong "breath, causeth an Earthquake. As is excellently, and very lively expressed "of Ovid, as I remember, thus:

Vis fera ventorum caecis inclusa cavernis" ....[3].

Compare Shakespeare's 1. Henry IV., Act III, I, 25—35:

Hotspur: "O, then the earth shook to see the heavens on fire,
"And not in fear of your nativity.
"Diseased nature oftentimes breaks forth
"In strange eruptions; oft the teeming earth
"Is with a kind of colic pinch'd and vex'd
"By the imprisoning of unruly wind
"Within her womb; which, for enlargement striving,
"Shakes the old beldam earth and topples down

---

[1] In de Jonge's "Opkomst", *I, 78*, it is stated that Jod. Hondius "*in Enge-"land* diversche globen over omtrent 10 jaren ende daernaer gesneden hadde". [I judge this to have been said in *1597*.] Hondius had also manufactured Molyneux's globes.

[2] "A Pleasant and Pitthy Familiar discourse, of the Earthquake in April last." pr. 1580.—Huth Libr. edition, Harvey, I, pp. 45 and 52.

[3] For parallel passages relating to the cause of earthquakes, see Parker Society "Liturgical Services", p. *570*: 'A Godly admonition', *1580*; Clarend. Pr. Ed. of 1. Henry IV., p. 145; Malone, XVI, p. 305.

"Steeples and moss-grown towers. At your birth
"Our grandam earth, having this distemperature,
"In passion shook."

And 'Venus and Adonis', 1046—'48:

"As when the wind, imprison'd in the ground,
"Struggling for passage, earth's foundations shakes,
"Which with cold terror doth men's minds confound."

ADDITIONAL NOTES.

1.—ADDENDA ET CORRIGENDA.

PAGE VIII. Regarding the booksellers' shops, see Ordish, Shakespeare's London, p. 233.

PAGE 5. A bibliographical account of the books and papers on Shakespeare is issued annually by the German Shakespeare Society in their Jahrbuch, to which Dr. Dibelius also contributes a 'Zeitschriftenschau', which will be found useful. The 'Shakespeare-Bibliographie' for 1900 and 1901 records the publication of more than 1280 books, articles, etc. relating to Shakespeare, and editions of his works. *Horresco referens.* Of course, in most cases the writers merely repeat old facts and ideas.[1]

PAGE 5 *note 1.* 'Macbeth' has now appeared in Liddell's edition, which he names the "Elizabethan Shakspere". Only 240 copies are published. Being expensive and somewhat unwieldy, it is not destined to displace the Globe Edition. We need a 'Shakespeare in Old Spelling', which, while absolutely reliable, must be cheap, handy, and practical for purposes of quotation and reference. The Globe edition excellent though it be, is by no means a textual *authority,* nor is its modernization always consistent. I give a few instances which occur to me at the present moment. In 1. Henry IV., Act I, III, the scene is laid at 'London, the palace'. This is not correct (see above, p. 253). In 'Romeo', II, IV, 151, Mercutio is made to *sing* "lady, lady, lady". This is mere conjecture and probably erroneous (cf. p. 166). Again, "the humour of forty fancies" is printed with quotation marks. These should be deleted (cf. p. 195). Why does the Globe edition read *Ætna* in Titus Andronicus (III, I, 242) but *Etna* in Merry Wives (III, v, 129); *Paul's* passim but *Powle's* in Henry VIII. (Act V, IV, 16);[2] *Cotswold* in 2. Henry IV.,

[1] All the English and German books and dissertations, etc., connected with English Literature and Philology ought to be indexed every year. A 'Poole's index' of the German learned journals is a sore want. Who will bring harmony into chaos? Of course, the best works should be marked with an asterisk.

[2] We find both spellings in Elizabethan books. Comp., e. g., the title page of the Quarto of the Merry Wives, 1602. But only *Paul's* should be admitted into a modernized text.

Act III, II, 23, but *Cotsall* in Merry Wives, I, I, 92; *Gloster* in 1. Henry VI., Act III, I, 49, but *Gloucester* elsewhere? Why is the reading *Bouciquall* (Henry V., Act III, v, 45) of the Folio replaced by *Bouciqualt* (as in Holinshed), but *Lestrale* not replaced by Holinshed's *Lestrake*?[1] On what principle are some passages and words in 2. and 3. Henry VI. adopted from the corresponding Quartos while others are rejected? Surely, the name *Hume* in 2. Henry VI. should be modernized into *Hum*[2]; surely, *lanthorn* (Mids. N. Dr.) should be spelt *lantern*; and for *Of it own kind* (Tempest, II, I, 163) we should read *Of its own kind*. (My Globe edition is dated 1895.) A close scrutiny of the Globe Edition would no doubt reveal many more little *errata*, such as I have pointed out.

PAGE 5 *note 1.* Regarding Fleay's Hist. of the Stage, cf. p. 119, note.
PAGE 10 *note 3*, *for* ,one *read* 'one
PAGE 15 *line 17*, for '*Vocavito* read '*Vocativo*
line 10 (from end), *for* he! *read* he![1]
note 1, *for* George Lyly's *read* John Lyly's
for where of *read* whereof
PAGE 16 *line 12*, for '*De* read *De*
PAGE 22 *note 1*, *for* ,Aureola *read* 'Aureola
PAGE 24 *line 21*, *for* noteworty *read* noteworthy
PAGE 26 *line 13*, *for* 'Never to Late' *read* 'Never too Late'
line 14 for 'Song *read* Song
note 1, compare Eckermann's Conversations with Goethe, Jan. 18 1825.
PAGE 28 *note 1.* But compare *post*, p. 113 *n* 4.
PAGE 29 *note 1* (last line), for *ut* read *(ut*
PAGE 33 *note 1.* 'Jack the Juggler' is founded on *Amphitruo.*
PAGE 36 *line 7*, *for* Bloud *read* 'Bloud
line 9, *for* say *read* 'say
line 13, *for* the Seneca *read* Seneca

PAGE 37 *lines 3 seq.* Mr. Hart, in his edition of Othello, 1903 (Arden Shakespeare), p. 161, quotes a later passage in Pliny (IV, 13): ....."out "of Pontus the sea alwaies floweth and never ebbeth againe". (Cf. also Hart, p. XXI.) Mr. P. A. Daniel called my attention to this.

PAGE 43 *line 8*, Mr. Fleay, in his Chronicle of the English Drama, II, 36, *says* that Dr. Legge wrote *The Destruction of Jerusalem,* and that this was acted at Coventry in 1577. I state this on Fleay's authority only.

PAGE 44 *line 6.* Mr. Churton Collins, in The Fortnightly Review, May, 1903, p. 848, disputes the priority of the discovery. He points out that

[1] Compare Boswell-Stone's "Shakspere's Holinshed", pp. 195—'6.
[2] Comp. Bosw.-Stone, p. 253 note 1.

"Dr. Wellesley in his *Anthologia Polyglotta* (1849), p. 93, printed sonnet CLIV., "without any remark, underneath the Greek original, as one of the versions." But Hertzberg made his discovery independently.

PAGE 44 *note 1 line 7, for* ‚Authour' *read* 'Authour'
PAGE 47 *line 2 from end, for* Comp, *read* Comp.,
note 1, for of the Bodleian *read* in the Bodleian
PAGE 48 *line 11 from end, for* they either o *read* they either of
PAGE 49 *line 10, for* II, 753, *read* II, 19, 753,

PAGE 51, I forgot to refer to Miss Hooker's article on Montaigne and Shakespeare, in the Publications of the Modern Language Association of America, Baltimore, 1902, pp. 312—366. She points out many 'parallelisms', many of which are doubtful. Mr. W. C. Hazlitt, too, gives us his opinion on the same subject, in his 'Shakespear' *(sic)*, 1902, pp. 155f.

PAGE 57 5), Perhaps Rabelais's influence is only indirect. We find similar fustian talk in Ben Jonson's Every Man *out* of his Humour, 1599, Act III, I.

PAGE 60 *note 4, for* wat *read* what
PAGE 63 *ll. 1—3,* Professor Gollancz made this suggestion, in his Temple Shakespeare, Cymbeline (Preface). But cf. note 3.

P. 65. Dr. Garnett tells me that recent discoveries among Egyptian papyri place the date of Chariton's work not later than the third century of our era.

PAGE 66 *note 1.* It was Steevens who made this remark regarding Titus and Painter. He can scarcely be supposed to have made a wilful error. Steevens wrote: "Painter, in his Palace of Pleasure, tom. II. speaks "of the story of Titus as well known, and particularly mentions the cruelty "of Tamora". (Malone, Var. Ed., XXI, p. 258.) Perhaps his mistake is due to confusion between *Timon* (referred to by Painter, vol. II) and Titus. In the Preface to the Reader *ibid.* Painter refers to "two Romayne Queenes" who "point (as it wer) with their fyngers, the natures of Ambitiō and Cruelty, "and the gredy lust (hidden in that feeble sexe) of soverainty". May we suppose that Steevens had made a hurried note regarding this, and that at a later date he rashly jumped to the conclusion that Tamora was referred to by Painter? Perhaps the name *Tanaquil* helped to add to the confusion. Not only truth but even error should be traced to its origin, if possible.

PAGE 71 *line 11f.* "Plautus", says Keller (Shakespeare Jahrbuch, XXXIV, p. 277), "hat, wie den Bramarbas so auch den Pedanten der neueren "Komödie zugeführt. Aus dem Ludus seiner *Bacchides* gehn alle Pädagogen "des italienischen Lustspiels hervor." Keller thinks that a Latin University play called 'Paedantius' (*circa* 1580?) was known to Shakespeare, which I very much doubt. Hermann Graf wrote a dissertation, Der Miles gloriosus im englischen Drama bis zur Zeit des Bürgerkrieges, 1891, Rostock-Schwerin. I have not seen this.

PAGE 75 *line 11, for* Tiek *read* Tieck

PAGE 77 *ll. 8—11.* The quotation is, of course, from Tennyson's *Dream of Fair Women.*

PAGE 78 *title, for* Dramatik *read* Dramatic

PAGE 81 *end.* Dr. Aldis Wright points out the following in *A World of Wonders* by Henry Stephen translated by R. C., 1607, p. 292: "the ancient Latinists .... had no good dexteritie in giving Etymologies of Ancient Latin words; witness the notation of *Mulier*, *quasi mollis aer*". A writer in *Notes and Queries* (Febr. 1857) quotes Isidore of Seville as giving this grotesque etymology. (I am quoting from Dowden's edition of Cymbeline, in the Arden Shakespeare, 1903, p. 209.)

PAGE 101 *note 2 line 1, for* Shakespeare *read* Shakespeare,

PAGE 104 *line 14*, for cayeth read 'cayeth

PAGE 108 *note 3.* *Breakespeare* (= Adrian IV), like *Shakespeare* and *Winspeare,* must be regarded as being imperatives (or perhaps infinitives) in form, which explanation is supported by names such as Makepeace, Do-well, Thudichum, Bleibtreu, Hauschild, etc., and words like forget-me-not, rendez-vous, portalettere (Ital.), etc.

PAGE 113 *line 3 from end.* Mr. P. A. Daniel doubted whether we should call Puck a fairy. But we know it from Puck's own mouth that he is one. Mids. N. Dr., V, I, 390: "*we* fairies", etc.

PAGE 115 *note 3 line 3, for* cru-/shed read crush-/ed

PAGE 127 *note 2.* Dr. Lessiak, in a work soon to be published, will prove that Dekker was the author of Lust's Dominion.

PAGE 128 *note 3.* Prof. Schick also edited The Spanish Tragedy with an introduction in The *Temple Dramatists*, 1898.

PAGE 134, *s. v. Mother Bombie.* I now prefer to regard the expression "I cry you mercy, I took you for a joint-stool" as a proverbial saying expressing a clumsy apology.

line 4 from end, for Shrew *read* Shrew,

PAGE 143 *note 4, dele* by the Meiningen Company. Kahle acted Timon in the *Schauspielhaus* (where now, alas, the chief *rôles* are played by Matkowsky, who appeals to the lower instincts of the audience).

Page 149 *1) and note 2.* Dr. Warner, Mr. W. W. Greg, and myself examined Henslowe's Diary most carefully with regard to the *ne*'s, on Sept. 8, 1903. If anybody's opinion is of weight, Dr. Warner's is. Mr. Greg, who is engaged upon a new edition of Henslowe's Diary, to be published next year by A. H. Bullen, is most familiar with the MS. volume. We used magnifying glasses and had good natural light. There can be no mistake whatever that the *ne*'s are authentic. Dr. Eichhoff's allegation that they were all forged by Collier is untrue, and he only discovered a mare's nest. Of the *ne*'s we examined, the only one which looked somewhat suspicious was that before 'titus and ondronicus', Jan. 23, 1593—4. But Dr. Warner was rather inclined to regard it as genuine.

Page 150 *note 2, for* it *read* the ballad

Page 154 *note 3.* I may here note one or two more *errata* in Schmidt's Lexicon: "*Oliver*", he says, is "a famous knight of Charlemagne's *Round* "*Table*". So far as I know, no Round Table is attributed to Charles. "*Claw* . . . . . used blunderingly: but age hath clawed me in his clutch, "Hml. V, i, 80. (Ff caught)". But clawed is in Tottel's Miscellany. *Baille* (Merry Wives, I, iv, 92) is not in the list of French words. Comp. also above, p. 168 and p. 257, p. 13f., p. 244 note 2; etc.

Page 155—195. I have already inserted some notes, with which the Rev. J. W. Ebsworth kindly favoured me, in the fifth chapter. I have much pleasure in subjoining the rest, which were crowded out above by the printer.

Page 156 *note 1 line 2.* J. W. E. conjectures that for tie "the true word "used was *tice* = entice. But we have no authority". — But tie gives very good sense.

Page 158 *line 3 from end.* Dr. R. Sievers is publishing a dissertation on Thomas Deloney, to appear in the 36th volume of the Palaestra, Berlin, 1903.

Page 164. J. W. E. refers me to his Robin Hood ballads, Roxburghe Ballads, vol. VIII (part xxiv), p. 529f, published in 1896. Child's third volume appeared in 1888/9.

Page 169 last lines. A contributor to the New Skakspere Soc. Trans., 1887—92, Part I, p. 142, says the words *calino casturame*=cailīn ōg as stuaire mē=a girl young and fair am I.

Page 170 *and passim.* J. W. E. tells me that I should not have quoted from, or referred to, Mr. Wooldridge's "garbled and emasculated" edition. The original edition of 1855 is "the only valuable edition". I admit Wooldridge's edition is inferior, but the older edition is out of print and Mr. Wooldridge gives some new information, though he earns the just condemnation of every scholar for the exclusion of most valuable matter. Nor would every scholar applaud his re-arrangement of the materials. As the original has a good index, which Mr. Wooldridge's edition has not, the reader will not be inconvenienced by my references to the latter. Of course, I used and studied both editions.

PAGE 173 *lines 4 and 5.* "Both ballads of Dr. Faustus and Titus An-"dronicus are reprinted in the Ballad Society, Roxburghe Ballads, vols VI, "703, and II, 544". J. W. E.

PAGE 173 *lines 16—24.* William Birch's ballad is in the Britwell Library. I applied to the trustees of the same for permission to copy this ballad. Mr. Graves replied: "I regret to have to inform you that it is not possible "to comply with your request because the publication of the ballad in question "would detract from the interest of a volume of early ballads and broadsides "in the same collection which may ere long be printed". (March 31st. 1903.)

PAGE 174 *Green Sleeves.* Chappell (orig. ed., p. 230) reprints, in part, the words given in 'A Handefull of pleasant delites'. I expressed my doubts to the Rev. J. W. Ebsworth whether these were the original words, as the Stat. Registers mention a newe *northern* Dittye of the Ladye Greene Sleves (3 Sept. 1580). In 1581 (13th Febr.) Elderton's ballad "A *Reprehension* againste Greene Sleves" was booked. J. W. E. sees "no reason to doubt that we have substantially the true ballad" as printed in "A Handefull of pleasant delites".

PAGE 181 *line 3.* 'Antidote against Melancholy' was reprinted by J. W. E. in 1876 in 'Choyce Drollery', etc.

PAGE 185. "Come away, come away, death" seemed to me to lack the ring of a truly popular song. It did not seem sufficiently 'silly' to me, nor to dally with the innocence of love. Hence I believed it was an interpolation. But the Rev. J. W. Ebsworth, whose opinion is better than mine, writes indignantly: "No, no, no! I disagree *in toto* with this guess and am confident "that it is Shakespeare's own intended song. I would stake my judgment "(not to say my reputation) on it. I maintain that it exactly fulfils the "premises,—precisely the pathetic song the spinsters and knitters would "sing at their work—and sung it."

PAGE 186. J. W. E. thinks my theory is 'doubtful'.—Some theory *must* be ventured to explain the very curious correspondence and resemblance between Ophelia's fragment, The Gentle Herdsman, and Deloney's ballad, which commences thus:

As you came from the holy-land
  Of Walsingham,
Met you not with my true love
  By the way as you came?

How should I know your true love,
  That have met many a one, *etc.*[1]

Then follows the description of the lady and the pilgrim's reply.—Bishop Percy must have instinctively felt the connexion, which subsists between the above productions, when he penned his 'Friar of Orders Gray'.

Either the original or some other Walsingham song is parodied in 'Hans Beer-Pot', 1618:

As I went to Walsingham,
  To that holy Land,
Met I with an olde balde Mare,
  By the way as I came.

For further information on Walsingham, see Chappell, orig. ed., pp. 121—'3. The original song is probably pre-Elizabethan.

PAGE 187 *lines 9f.* The Rev. J. W. Ebsworth had in mind "a ballett intituled *taken nappynge as Mosse toke his meare*", which we find entered on the Stationers' Registers in 1569/70 (Arber, I, 417).

PAGE 187 *line 29.* 'This is not certain', I said. But there can be no reasonable doubt about this any more, as the Rev. J. W. E. points out to me. For there is a better version than the one printed by Child from the Percy MS. in the *Choyce Drollery*, 1656, republished by J. W. Ebsworth in 1876 (Boston, Lincolnshire). The ballad (pp. 78—80) is entitled 'Upon the Scots being beaten at Muscleborough field', and begins thus:

On the *twelfth*[2] day of *December*,
In the fourth year of King *Edwards* reign[,]
Two mighty Hosts (as I remember)
At *Muscleborough* did pitch on a Plain.
For a down, down, derry derry down, Hey down a,
Down, down, down a down derry.

PAGE 189 *line 7 from end.* 'But I suspect Collier is swindling'. To this J. W. E. remarks. "This to me is vile and gratuitous insult against the "dead man, whom I love and honour, and knew for years."

---

[1] Percy Soc., XXX, Garland of Goodwill, p. 111. (The G. of G. is referred to by Nash, in *Have with you,* 1596, Huth Libr. ed., p. 123. As to *Walsingham* cf. *ibid,* p. 98.)

[2] My italics.

PAGE 191 *line 9.* Shakspere. "*Never* yourself tolerate this inaccurate "spelling: brought in culpably by Chas. Knight in his Pict. Shaksp., and "imitated by others." J. W. E.

PAGE 193 *line 3, for* "The Merry Wives" *read* "The Merry Wives of Windsor".

PAGE 200 *line 3, for* bollie *read* bellie

PAGE 201. In The Athenaeum, April 28, '83, p. 542, Mr. Ginsburg points out that the following passage in Love's Labour's Lost, IV, III, 364-'5:

"For *charity* itself fulfils the law,
"And who can sever love from charity?"

comes nearest to the translation of *Romans*, XIII, 10, in the Bishops' Bible: 'Therefore the fulfyllyng of the lawe is *charitie*', which the other bibles render: 'Therefore is *love* the fulfilling of the law.'

PAGE 217. Metrical Psalms are still sung in Scotland.

PAGE 225 *line 16,* 'the first that ever burst into that silent sea' is, of course, from Coleridge's *Ancient Mariner* (Part II, st. 5).

PAGE 225 *line 19.* On *Setebos* Mr. H. Hesketh Prichard (at Southsea) the author of *Through the Heart of Patagonia*, London, 1902, very courteously wrote the following lines to me: "I will answer your letter in full in a few days. I am writing to a friend, who has lived among the Tehuelche his childhood, for an opinion to see if it corroborates yours." Again, on Aug. 17. 1903: "I have never heard the Tehuelche Indians of Patagonia "mention *Setebos.* As far as I could gather, their religious beliefs only "recognise a good spirit which is nameless and an evil one — the Gualicho. "This is endorsed by my correspondent whom I mentioned in my last letter."

PAGE 239 *line 10.* Some remarks on Shakespeare's astrology will be found in MacMillan's Magazine, 51, p. 462.
last lines. But compare Sonnet XV, 3 and 4.

PAGES 250—3. To the London names add the following: *The Jewel House* of the Tower (mentioned in Henry VIII., Act IV, I, 111, and V, I, 34); **Long Lane* (? The Shrew, IV, III, 187); **the pissing conduit* (2. Henry VI., Act IV, VI, 3); **The Tiltyard* (2. Henry IV., Act III, II, 347).

2.—MORE BORROWED IDEAS.

> Nihil novum est sub sole; nihilque dictum, quod non sit dictum prius.

I now offer a few jottings on further Shakespearean borrowings. I say jottings, for I am reluctantly compelled by circumstances (not by space this time) to give up my intention of writing a separate chapter on this subject. Fragmentary though this section be, there is perhaps some gain in at least stating what has still remained undone. The chapter was to include:— 1. More classical elements in Shakespeare's works. This I have partly worked out. 2. Ideas of which parallels can be adduced from contemporary or earlier literature. 3. Proverbs and sayings. 4. Folk-lore in Shakespeare; and Popular Errors regarding animals, plants, etc. 5. Inquiry into Shakespeare's knowledge of law, music, medicine, etc.

There is no saying where we should stop our investigation. For is not the very language which Shakespeare spoke the work of generations moulded and refined by many varied minds?

1.—More Classical Elements in Shakespeare's Works with many other Parallels.

In this section, which is supplementary to chapter 1, I shall mention, by way of example, some ideas and expressions which can be traced back to the old classics. It is not always possible to say exactly whether the influence is of a direct or indirect kind. But in many cases we feel certain that it is of the latter. For the whole Renaissance atmosphere was surcharged with classicism, and one who had enjoyed a very fair classical schooling himself and who moved among the best of his age would continually add to the store of his knowledge.

The question of Shakespeare's relation to the classics has lost nothing of its fascination. A suggestive paper on Shakespeare's Predecessors was read by Professor Brandl before the German Shakespeare Society at Weimar in 1899 (Jahrbuch, xxxv). He shows that, though Shakespeare cannot be supposed to have studied the Greeks, there is a plain chain of connexion stretching across the centuries from the Elizabethan to the Attic stage. The

Fortnightly Review, as I have already stated, brings a series of articles to prove that Shakespeare had read the Greek dramas in Latin versions. Mr. Churton Collins, the author of these articles, was no doubt influenced by Mr. Edwin Reed's chapter on classic elements in Shakespeare's plays, in his "Francis Bacon our Shake-speare" (London, 1902), a work which cannot stand the test of serious criticism. It is only his parallelisms between Shakespeare and the classics which we are concerned with here. A detailed discussion of his and Mr. Collins's remarks is, of course, out of question. I can only select more noteworthy points, which I shall refer to *en passant*. Mr. Collins's position I may remark, is not new. He had been anticipated by James Russell Lowell.[1]

I confess, I have much sympathy with those, who, impressed with the complexity of the gigantic mind of Shakespeare, like to think that he could have been little satisfied with second-hand knowledge. For him only the best was good. It is, indeed, difficult to draw a hard and fast line between direct and indirect influences. No doubt, Shakespeare was a hard-working man. Supposing he spent an hour every morning before breakfast in reading belletristic literature, why should he not have opened the Latin writings now and then? If so, why not the Greek authors in the Latin (or French) translations? But there is a limit to everything,—except to scientific enquiry.

I have already said in Chapter 1 that Shakespeare *did* know the Latin classics in the original. But, though they supplied him with a sound training, their influence should not be exaggerated. The very fact that he opened his Golding should be a sufficient warning. It is of course *possible* that Shakespeare had looked into the Greek classics in the Latin translations. There are many things about which *ignoramus et ignorabimus*.

What has Mr. Collins shown? He has displayed much learning and research and proved that he himself has carefully studied the Greek dramas; but he has conspicuously failed to show Shakespeare's obligations to them, so much so, that his articles may be cited as providing strong evidence to the contrary. Even supposing that, by some chance, English translations of the Greek dramas should be discovered (which is extremely unlikely), I should continue to remain sceptical as to whether Shakespeare studied them, basing my doubts on the inconclusiveness of Mr. Collins's articles. His paper is full of honest destructive self-criticism, which—it is not offensive to say so—shows that Mr. Collins has failed to convince himself. The difference between Shakespeare's drama and the Attic tragedy is essential and radical. Shakespeare's mind travelled in another direction altogether, and the connexion, such as it is, is indirect through the Elizabethan contemporary drama and literature at large. For a fair judgment on this question a constant com-

[1] Compare Mr. Lee's Life of Shakespeare, p. 13*n*. "I believe", says Mr. Lee, "Lowell's parallelisms to be no more than curious accidents—proofs of consanguinity of spirit, not of any indebtedness on Shakespeare's part."

parison with the Elizabethan literature is imperative,—a comparison which Mr. Collins has evaded entirely. The subject, if investigated again, must be approached from a standpoint offering a broader and wider view.[1]

Need we repeat, what is a truism, that the language of Nature is the same in London as in Athens; that the Attic Muse had been resuscitated by the Renaissance; and that the basis of Shakespeare's drama is national not exotic?

I begin by pointing out some thoughts and expressions, which can be traced back to the old Philosophy, and which must have been familiar to the poet through the Renaissance writings.

1) *I do now remember a saying, 'The fool doth think he is wise, but the wise man knows himself to be a fool'.*

(As You Like It, V, I, 33.)

This is a Socratic idea.

2) *Look, who comes here!* a grave unto a soul;
Holding the eternal spirit, against her will,
In the vile prison *of afflicted breath.*

(King John, III, IV, 17—19.)

Now my soul's palace is become a prison, *etc.*

(3. Henry VI, Act II, I, 74—'6.)

this hollow prison *of my flesh.*

(Titus, III, II, 10.)

Compare what Plato says, e. g., in his 'Cratylus', 400: 'For some say that 'the body is the grave (σῆμα) of the soul which may be thought to be buried 'in our present life; or again the index of the soul, because the soul gives 'indications to (σημαίνει) the body; probably the Orphic poets were the in- 'ventors of the name, and they were under the impression that the soul is 'suffering the punishment of sin, and that the body is an enclosure or pri- 'son in which the soul is incarcerated, kept safe (σῶμα, σώζηται), as the name 'σῶμα implies, until the penalty is paid'.[2] The thought had no doubt become a commonplace.

3) *To be, or not to be . . . . .*
To die: to sleep;
No more; and by a sleep to say we end
The heart-ache and the thousand natural shocks
That flesh is heir to, 'tis a consummation
Devoutly to be wish'd. To die, to sleep;

---

[1] Two somewhat puzzling allusions in Titus Andronicus are noted below, p. 285.

[2] Plato, translated by Prof. Jowett, 1892.

To sleep: perchance to dream: *ay, there's the rub;*
For in that sleep of death what dreams may come, etc.
(Hamlet, III, I, 56f.)

'Let us reflect', says Socrates in Plato's *Apology* (40), in another way, 'and we shall see that there is great reason to hope that death is a good; 'for one of two things—either death is a state of nothingness and utter un- 'consciousness, or, as men say, there is a change and migration of the soul 'from this world to another. Now if you suppose that there is no conscious- 'ness, but a *sleep* like the sleep of him who is undisturbed even by *dreams,* 'death will be an unspeakable gain . . . . . Now if death be of such a nature, 'I say that to die is gain; for eternity is then only a single night.[1] But if 'death is the *journey* (ἀποδημῆσαι) to another place, and there, as men say, 'all the dead abide, what good, O my friends and judges, can be greater 'than this?'

The passage, from which these sentences are taken, was very famous in antiquity. It is translated by Cicero (Tusc. Disp. I, 97—99), and quoted by Stobaeus and Eusebius: Plutarch also gives a paraphrase of the reasoning in his Consolatio ad Apollonium, 107 D foll. (cf. Holland's transl. of the "Morals", 1603, p. 516). Compare also Xen. Cyrop. VIII. 7. 19 foll., where the dying Cyrus talks much as Socrates does here.[2] Montaigne repeats Socrates's thoughts (cf. p. 54), of which we also find an echo in Ph. de Mornay's Discourse of Life and Death, translated by Mary, Countess of Pembroke, 1592 (2nd ed. 1600). With the above passage we should also compare Measure for Measure, III, I, 17:

"Thy best of rest is sleep,
"And that thou oft provokest; yet grossly fear'st
"Thy death, which is no more." etc.

or The Tempest, IV, I, 156—8:

"We are such stuff
"As dreams are made on, and our little life
"Is rounded with a sleep."

4) *The beast*
With many heads butts me away.
(Coriol., IV, I, 1.)

The blunt monster with uncounted heads,
The still-discordant wavering multitude.
(2. Henry IV., Ind. 18.)

---

[1] Huxley's tombstone has the following inscription: It is well even if the sleep be endless.

[2] Adam's edition of the Apology, 1887, p. 115.

In the Republic (IX, 588) the human soul is compared to a multitudinous many-headed monster. The Stoic Ariston of Chios calls the people a πολυκέφαλον θηρίον. Hence Horace's *belua multorum capitum* (Epist., I, I, 76). Stephen Gosson, in his *Plays Confuted*[1] (1582) writes: 'the auncient Philo-'sophers . . . called them [the people] a monster of many heades'. Compare, also, Daniel's Complaint of Rosamond (v. 279): 'many headed beast'. Again, Hakluyt's Voyages. 1599, vol. II, part 2, p. 142: 'shall the blind opinion of 'this monster, a beast of many heads, (for so hath the generalitie of old bene 'termed) cause me to neglect the profession?' etc.—I refrain from quoting other instances.

5) *Come hither. If thou be'st valiant,—as*, they say, *base men being in love have then a nobility in their natures more than is native to them,—list me.*

(Othello, II, I, 216—'8.)

Compare Plato's Symposium, 179: 'The veriest coward would become 'an inspired hero, equal to the bravest . . . Love would inspire him', etc.[2]

6) *Virtue is beauty*

(Twelfth Night, III, IV, 403.)

The germ of this is, I believe, to be found in Plato. Compare, e. g., 'Symposium', 201, *Socr.* τἀγαθὰ οὐ καὶ καλὰ δοκεῖ σοι εἶναι *Ag.* Ἔμοιγε. Bishop Wordsworth, in his work on Shakespeare's Knowledge and Use of the Bible, 1880, p. 218, considers this idea to be derived from the stoical philosophy. Be this as it may, the thought was familiar to the Elizabethans. Sidney, in his 5th Sonnet of 'Astrophel and Stella' says: "True—that true beauty Virtue "is indeed." Or compare Spenser's 'Hymn in Honour of Beauty', where we read: "For all that fair is, is by nature good" (l. 139).[3]

7) In Troilus and Cressida, III, III, 95 *seq.* we have the following passage:

Achil. *What are you reading?*
Ulyss. *A strange fellow here*

---

[1] Hazlitt, The Roxb. Libr., Engl. Drama and Stage, 1869, p. 184.

[2] Platonic ideas in Shakespeare's sonnets are pointed out by H. Isaac, in Archiv, 61, p. 193f.

[3] Compare with Spenser's poem the entire passage in Twelfth Night, III, IV, 399—404. Also The Tempest, I, II, 457:

There's nothing ill can dwell in such a temple:
If the ill spirit have so fair a house,
Good things will strive to dwell with't.

In Arthur Brooke's Romeus and Juliet (l. 406) we read:—

in no wise can it be,
That where such perfect shape, with pleasant beauty rests,
There crooked craft and treason black should be appointed guests.

Writes me: 'That man, how dearly ever parted,
'How much in having, or without or in,
'Cannot make boast to have that which he hath,
'Nor feels not what he owes, but by reflection;
'As when his virtues shining upon others
'Heat them and they retort that heat again
'To the first giver.'
Achill. *This is not strange, Ulysses.*
The beauty that is borne here in the face
The bearer knows not, but [=unless?] commends itself
To others' eyes; nor doth the eye itself,
That most pure spirit of sense, behold itself,
Not going from itself; but eye to eye opposed
Salutes each other with each other's form;
For speculation [=vision] turns not to itself,
Till it hath travell'd and is mirror'd there
Where it may see itself. This is not strange at all.
Ulyss. *I do not strain at the position [=assertion],—*
It is familiar,—but at the author's drift;
Who, in his circumstance [=deduction], expressly proves
That no man is the lord of any thing,
Though in and of him there be much consisting,[1]
Till he communicate his parts to others,
Nor doth he of himself know them for aught
Till he behold them form'd in the applause
Where they're extended: who, like an arch, reverberates
The voice again, or, like a gate of steel
Fronting the sun, receives and renders back
His figure and his heat.

I quote this passage entirely, because Mr. Churton Collins, in his article "Had Shakespeare read the Greek tragedies?", in the so-called Fortnightly Review, April, 1903, p. 632, finds here "conclusive proof" of Shakespeare's acquaintance with the Greek classics in Latin versions. He compares the pseudo-Platonic dialogue *Alcibiades I*, p. 132, 133, where we find the following curious argument:

As the eye sees itself by looking at the most perfect part, i. e. the pupil, of the eye of another, so the soul which would know herself must look at the most perfect part of herself, that is to say, at that part which has to do with wisdom and knowledge:

Socr. Did you ever observe that the face of the person looking into the eye of another is reflected as in a mirror; and in the visual organ

[1] =though much may depend on his cooperation and power.

which is over against him, and which is called the pupil, there is a sort of image of the person looking?[1]

Alc. That is quite true.

Soc. Then the eye, looking at another eye, and at that in the eye which is most perfect, and which is the instrument of vision, will there see itself?

Alc. That is evident.

Soc. But looking at anything else either in man or in the world,[2] and not to what resembles this, it will not see itself? etc.

It will be observed that the *drift* of the "author" whom Ulysses is reading is quite different. What the passage in the play means is this. Man has what he has and knows what he has by reflection. Just as beauty knows not itself without commending itself to others, and as the eye is mirrored in another eye, so a man's gifts must be imparted and he only becomes conscious of them by the applause of those to whom they are directed.

It is plain that Mr. Churton Collins (cf. *ante*, p. 44) has exaggerated the degree of resemblance. So far from concurring with him in the assertion that here we find a "conclusive proof" of Shakespeare's familiarity with the Greek classics, I am rather inclined to put the question whether the resemblance is not altogether accidental.

8) *For government, though high and low and lower,*
Put into parts, doth keep in one consent,
Congreeing in a full and natural close,
Like music.

(Henry V., Act I, II, 180.)

A remarkable parallelism to this was pointed out by Theobald from Cicero's *Republic*, of which we possess only fragments. The passage, which was known to the Elizabethans only from St. Augustine's *De Civitate Dei*, II, 21, is this:

'Cum autem Scipio in secundi libri fine dixisset, ut in fidibus ac tibiis 'atque cantu ipso ac vocibus concentus est quidam tenendus ex distinctis 'sonis, quem immutatum aut discrepantem aures eruditae ferro non possunt, 'isque concentus ex dissimillimarum vocum moderatione concors tamen efficitur 'et *congruens*; sic ex *summis et infimis et mediis* interjectis ordinibus, *ut sonis*, 'moderata ratione civitatem consensu dissimillimorum *concinere*; et quae 'harmonia a musicis dicitur in cantu, eam esse in civitate concordiam; arc- 'tissimum atque optimum omni in republica vinculum incolumitatis, eamque 'sine justitia nullo pacto esse posse.'

---

[1] Comp. Jul. Caesar, I, II, 52f.

[2] Excepting a mirror!

Cicero seems to have been indebted to Plato's Republic (IV, 432) for this idea. Dr. Aldis Wright thinks that Shakespeare borrowed it from no one.

9) To the theory of the transmigration of the souls held by the Pythagoreans, we find allusions in Twelfth Night, IV, II, 54f., in As You Like It, III, II, 187—'8, and Merchant of Ven., IV, I, 131f.

Besides mentioning the names of Pythagoras, Aristotle, Epicurus, and Socrates, Shakespeare also alludes to Heraclitus, though without mentioning his name, in The Merchant of Venice, I, II, 52f.:

'*I fear he will prove the weeping philosopher.*'

10) The philosopher Anaxarchus is said to have bitten off his tongue and spit it into the face of the tyrant who tortured him. Sir Thomas Elyot, in his work 'Of that knowlage, whiche maketh a wise man', 1534, Dialogue 5, attributes this to Zeno Eleates. This anecdote was known to Shakespeare. Compare Richard II., Act I, I, 190—'5.

We now proceed to point out some thoughts and expressions in Shakespeare, for which we can find instances and parallels in miscellaneous writings of the antique literature.

11) *All the world's a stage, etc.*
(As You Like It, II, VII, 139; cf., too Merch. of Ven., I, I, 77—'8.)

See Furness's note in his Variorum Edition, VIII, 121; and Schelling, Elizabethan Lyrics, p. 263.[1] The sentiment was very common in Shakespeare's time. *Totus mundus agit histrionem* was the motto to the Globe Theatre.

12) Man's Life was divided into *seven ages* (see As You Like It, *ut sup.*, 143f.) already by Hippocrates, Proclus, and others. See Furness, VIII, 122f., and Notes and Queries, 9th S., IX, 46, etc. Ralegh's History of the World, printed in 1614, contains the same subdivision. (Compare Ward, Hist. of Engl. Drama, II, 131—'2; Arber's Engl. Garner, I, 139; Staunton's Shakespeare, vol. II, 466.)

13) *For to be wise and love*
Exceeds man's might; that dwells with gods above.
(Troilus and Cress., III, II, 163.)

Compare 'amare et sapere vix deo conceditur', a sentence ascribed to Publius Syrus. This was translated thus by Spenser in his Shepherd's Calendar, March:

To be wise and eke to love
Is graunted scarce to gods above.

---

[1] Comp. New Shaksp. Soc. Trans. 1880—6, p. 9* and 10*.

A similar passage is also in Marston's Dutch Courtezan.

14) *Sleep that knits up the ravell'd sleave of care,*
The death of each day's life, sore labour's bath,
Balm of hurt minds, great nature's second course,
Chief nourisher in life's feast.

(Macbeth, II, II, 37.)

Compare Ovid, *Met.* XI, 623:

Somne, quies rerum, placidissime Somne Deorum,
Pax animi, quem cura fugit; qui corda diurnis
Fessa ministeriis mulces, reparasque labori.[1]

Also *Met.*, VIII, 81 f.:

Talia dicenti, curarum maxima nutrix
Nox intervenit; tenebrisque audacia crevit.
Prima quies aderat; qua curis fessa diurnis
Pectora somnus habet—

and Virgil, *Aen.* IV, 527:

somno positae sub nocte silenti
Lenibant curas, et corda oblita laborum.[1]

The following passage in Sidney's Astrophel and Stella (Poem XXXIX) was, of course, known to Shakespeare:

Come, Sleepe! O Sleepe, the certaine knot of peace,
The *bathing*[2] place of wits, the *balme* of woe,
The poor man's wealth, the prisoner's release, etc.
Th' indifferent iudge betweene the high and low.

Compare also Capell's note in Furness's Var. Ed.

15) *We came crying hither:*
Thou know'st, the first time that we smell the air,
We wawl and cry.

(Lear, IV, VI, 182—'4.)

I have already pointed out a curious coincidence with Holland's Pliny (cf. *ante*, p. 37). The thought may also be found in Lodge's and Sidney's works (cf. Notes and Queries, 9th S., vol. IX, pp. 197 and 298). Compare also Tim. Kendall, Trifles appended to his 'Flowers of Epigrammes', 1577 (Spenser Soc. ed., p. 10):

---

[1] Compare also Seneca, Hercules Furens, 1065 ff.
[2] 'baiting-place' is the reading of the edition of 1598.

Of the misery of man.
We weping come into the world: and weping hence we goe,
And all our life is nothyng else, but grief, payne, toyle, and wo.

16) *Look like the innocent flower,*
But be the serpent under 't.

(Macb., I, v, 66.)

We find the snake under the flower again in Richard II., Act III, i, 19—22; in 2. Henry VI., Act III, I, 228—'30; compare, too, Romeo, III, ii, 73. The idea is in Virgil's *Eclogue,* III, 92: 'Qui legitis flores . . . fugite hinc, 'latet anguis in herba.' Arthur Brooke in his Romeus and Juliet (l. 386) speaks of 'the snake that lurks in the grass'. Again, in Byrd's 'Psalms Sonnets and Songs', 1588, we read: 'a poison'd serpent cover'd all with 'flowers'.[1] Gascoigne, too, knows that 'In sweetest flowres the subtyll Snakes 'may lurke'.[2] Compare also Chaucer, Squire's Tale, vv 504 ff. The idea was, of course, seized upon by the emblematists and moralists (cf. H. Green, Shakespeare and the Emblem Writers, p. 340).

17) *Nor marble, nor the gilded monuments*
Of princes, shall outlive this powerful rhyme.

(Sonnet 55.)

This thought which recurs in many of Shakespeare's sonnets, is, of course, derived from Horace's famous ode *Exegi monumentum aere perennius* (Ode, III, 30) or from Ovid: *Jamque opus exegi, quod nec Jovis ira nec ignes Nec poterit ferrum, nec edax abolere vetustas* (Met., XV, 871—'9). But it had become the brag of numberless poets and rhymesters. Compare Sidney Lee, Life of Shakespeare, pp. 114 and 116.

18) *Were she as rough*
As are the swelling Adriatic seas.

(The Shrew, I, ii, 73.)

Compare Horace's *Improbo iracundior Adria* (Ode, III, ix, 22) or his *fretis acrior Adriae* (Ode, I, xxxiv, 15).

19) The *"fine frenzy"*

of the poet (Mids. N. Dr., V, i, 12) has been connected with Horace's *amabilis insania* (Ode, III, iv, 5). When we have to deal with a mind like Shakespeare's, as wealthy in imagery and invention as assimilative in faculty, it becomes at times difficult to decide to any degree of confidence the claims which may be put forward on the side either of accidental coincidence or of influence.

---

[1] A. H. Bullen, Lyrics from Elizabethan Song-books, p. 24.
[2] Gascoigne's Poems ed. by Hazlitt, 1870, vol. II, p. 110.

20) *The undiscover'd country from whose bourn*
No traveller returns.

(Hamlet, III, I, 79.)

Steevens quoted from Catullus (*Carm.*, III, 11):

Qui nunc it per iter tenebricosum
Illuc, unde negant redire quenquam.

But we meet with this idea also in Anacreon, who writes: "Because "the dungeon of Acheron is dismal and there is no return for anyone who "has descended into it". Theocritus (12, 19) also says ἀνέξοδον εἰς Ἀχέροντα; and Philetas: Ἀτραπὸν εἰς Ἀΐδεω ἤνυσα, τὴν οὔπω τις ἐναντίον ἦλθεν ὁδίτης (= Iter ad inferos feci, per quod nullus retrorsum viator venit). But Marlowe's Edward II., Act V, VI, 64:

Mort. Farewell, fair queen; weep not for Mortimer,
That scorns the world, and, as a *traveller,*
Goes to *discover countries* yet unknown.

was known to Shakespeare. A careful search through the Elizabethan writings would probably reveal many more parallelisms in this as in other cases. 'The same idea is found in Sandford's version (1569) of Cornelius Agrippa'.

21) *Men's evil manners live in brass; their virtues*
We write in water.

(Henry VIII., Act IV, II, 45.)

Catullus (Carm. LXX) says:

Dicit: sed mulier cupido quod dicit amanti,
In vento, et rapida *scribere* opertet *aqua.*

This was translated thus by Sidney

These be her words; but a woman's words to a love that is eager,
In wind or waters streame do require to be writ.[1]

Again, in 'Philaster' by Beaumont and Fletcher (who is the probable collaborateur of Henry VIII.) we read:

All your better deeds
Shall be *in water writ*, but this in marble.

22) *Thou art an* elm, *my husband, I a* vine, etc.

(Com. of Err., II, II, 176).

Sleep thou, and I will wind thee in my arms.

---

[1] Compare Grosart's edition of Sidney, vol. II, pp. 43—44,

Fairies, be gone, and be all ways away.
So doth the woodbine the sweet honeysuckle
Gently entwist; the female ivy *so*
Enrings the barky fingers of the elm.
(Mids. N. Dr., IV, I, 43 f.)

Catullus (LXII, 54) says of the vine.

ulmo conjuncta marito;

and in Ovid (Met., IV, 365) Salmacis embraces Hermaphroditus,

utve solent hederae longos intexere truncos.

It is easy to give parallel passages from Elizabethan authors. Daniel, in his Complaint of Rosamond, says:

And as the Vine married unto the Elme
With strict embraces, so doth he infold it [the body].[1]

In Kyd's Spanish Tragedy, Act II, II, we read:

Bell. My twining arms shall yoke and make thee yield.
Hor. Nay then my arms are large and strong withal:
Thus elms by vines are compass'd till they fall.

Again, in Sidney's Arcadia we find the following verses:

The honest Bridegroome and the bashfull Bride,
Whose loves may ever bide
Like to the *elme* and *vine*
With mutuall embracements them to twyne.[2]

Lastly, in Chaucer's Troilus and Cress., III, l. 1230 we read:—

And as aboute a tree with many a twiste,
Bitrent [= wind] and wryth the swote *wode-binde*,
Gan eche of hem in armes other winde.

Compare also Green, Shakespeare and the Emblem Writers, p. 308.

23) *The earth that's nature's mother is her tomb;*
What is her burying grave that is her womb.
(Romeo, II, III, 9—10.)

Steevens cites Lucretius, *De Rer. Nat.*, V, 260:

Omniparens eadem rerum commune sepulcrum.

But Shakespeare need not have gone to Lucretius; though, of course, it is no heresy to suppose he did.

---

[1] Grosart's edition, p. 110.
[2] Sidney, Poems ed. by Grosart, III, 15.

24) *Lay her i' the earth:*
And from her fair and unpolluted flesh
May violets spring.

(Hamlet, V, I, 261.)

Compare Persius (*Sat.* I, 38):

Nunc non e manibus illis,
Nunc non e tumulo fortunaque favilla
Nascentur violae?

25) *The two bears will not bite one another when they meet.*

(Much Ado, III, II, 80.)

One bear will not bite another.

(Troilus & Cress, V, VII, 19.)

Compare Juvenal's *saevis inter se convenit ursis* (Sat. XV, 164). Perhaps it had become proverbial.

26) *Will all great Neptune's ocean wash this blood*
Clean from my hand?

(Macbeth, II, II, 60.)

Parallels have been quoted from Seneca[1], Catullus, Sophocles, from Hall's Satires, etc. Pilate also washed the blood off his conscience.

27) *He seem'd in running to devour the way.*

(2. Henry IV., Act I, I, 47.)

Compare Catullus, XXXV, 7:

Quare, si sapiet, *viam vorabit.*

The expression 'devour the way' is also used by Ben Jonson, Sejanus, Act V. sc. x. Compare also the French *dévorer l'espace,* or *le temps.*

28) *To make a virtue of necessity.*

(Gent. of Ver., IV, I, 62.)

'Fac de necessitate virtutem', says Hieronymus.[2] 'To maken vertue of necesitee', is in Chaucer's Knight's Tale, A. 3042. But we are already in the sphere of proverbs.

29) *a sea of troubles.*

(Hamlet, III, I, 59.)

a sea of care.

(Lucr. 1100.)

Compare the Greek κακῶν θάλασσα, κακῶν πέλαγος. Marlowe uses a similar expression 'a sea of tears' (2. Tamburl., III, II, 47). In the 'Mirror for

---

[1] Further parallelisms between Shakespeare and Seneca are given by Cunliffe, in his worked quoted above, p. 34 *n* 1.

[2] Comp. Büchmann, Geflügelte Worte.

Magistrates', in The Tragedy of Queen Cordilla we read: 'to tell my seas of giltlesse smart'. (Compare the Variorum Editions.)

30) *Devouring time.*

(Sonn. XIX, 1.)

Tooth of time

(Meas. f. M., V, I, 12.)

Simonides of Ceos (*d.* 468 B. C.) spoke of "sharp-toothed time". Compare also Ovid, Metam., XV, 234:

> *Tempus edax* rerum, tuque, invidiosa vetustas,
> Omnia destruitis: vitiataque *dentibus aevi*
> Paulatim lenta consumitis omnia morte.

We now take note of two rather puzzling passages in Titus Andronicus.

31) *The self-same gods that arm'd the Queen of Troy*
With opportunity of sharp revenge
Upon the Thracian tyrant in his tent,
May favour Tamora, the Queen of Goths.

(Act I, I, 136—9.)

Theobald proposed the reading "in *her* tent",—'i. e. in the tent where she 'and the other Trojan captive women were kept: for thither Hecuba by a 'wile had decoyed Polymestor, in order to perpetrate her revenge. This we 'may learn from Euripides's *Hecuba*; the only author, that I can at present 'remember, from whom our writer must have gleaned this circumstance'. (Malone, Var. Ed., XXI, 269.) Ovid, in his Metamorphoses, XIII, recounts how Hecuba wreaked vengeance on Polymestor, the king of Thracia, but he says nothing about her tent.

The second passage is in the same scene of the same play (ll. 379—381).

32) *The Greeks upon advice did bury Ajax*
That slew himself; and wise Laertes' son
Did graciously plead for his funerals.[1]

Steevens observed: 'we have here a plain allusion to the Ajax of Sophocles, of which no translation was extant in the time of Shakespeare. In 'that piece, Agamemnon consents at last to allow Ajax the rites of sepulture, 'and Ulysses ['Laertes' son'] is the pleader, whose arguments prevail in favour 'of his remains'. (Malone Var. Ed. XXI, p. 280.)

What explanation shall we offer for these two allusions? Shall we accept Theobald's and Steevens's explanation; and shall we say that here we have

[1] Compare also 2. Henry VI., Act V, I, 26:
And now, like Ajax Telamonius,
On sheep or oxen could I spend my fury.

signs of another academic hand in Titus Andronicus? Or shall we suppose that Shakespeare was made thoroughly familiar with the Trojan story by his Stratford teacher, who would have told him more about Ajax and Hecuba when reading the thirteenth book of the Metamorphoses? Or did Shakespeare gain his knowledge in another way? from plays on Troy and Ajax?

33) The striking resemblance of the closing scene of The Winter's Tale, where Hermione reappears as a statue, to the last of Euripides's *Alcestis* has often been noticed. I have no doubt but that the story of Admetus and Alcestis was known to Shakespeare.

Mr. Edwin Reed opens his book on Bacon and Shakespeare, 1902, by producing what he considers "nearly absolute proof" that the author of Henry V. (of course Bacon) had studied Cardan's translation of Hippocrates's *Prognostica*, or Galen's commentary upon it. I have carefully investigated the matter and consider his arguments baseless.

Mr. Collins, in The Fortnightly Review, April, 1903, p. 635, says that Shakespeare's "*rifted* Jove's stout *oak*" accurately *(sic)* recalls Ovid's "suā convulsaque robora terrā",—which needs no comment. The other coincidences also regarding Tempest, V, I, 33 f., which he points out on the same page seem to me altogether unimportant.

'The Nation' (New York, 1875, 11 March)[1] asks whether the following touch in 1. Henry IV., II, III, 90,

In faith, I'll break thy little finger, Harry,
An if thou wilt not tell me all things true—

is not taken from Polydore Vergil. I have wasted about three hours in trying to discover the passage in question, but can find no trace of it. The reference in the 'Nation' is as irritating as it is useless. Dr. W. Aldis Wright, in his Clarendon Press edition, p. 128 says: "*I'll break thy little finger.* A reference "to an old custom not yet forgotten".

II.—Passages with Parallels from Modern Literature.

1) *Come on, come on; you are pictures out of doors,*
Bells in your parlours, wild-cats in your kitchens,
Saints in your injuries, devils being offended,
Players in your housewifery, and housewives in your beds.
(Othello, II, I, 110 f.)

In Middleton's 'Blurt, Master Constable', printed 1602, we find a similar passage (Act III., III):

[1] Compare New Shaksp. Soc. Trans., 1875/76, p. 124; and p. 462, line 10. Perhaps we are fooled by a forger.

according to that wise saying of you, be saints in the church, angels in the street, devils in the kitchen, and apes in your bed.

A very close parallel to this occurs in Puttenham's, *Arte of Poesie* (p. 299, Arber's ed.).

Compare, too, Florio, Second Frutes, 1591, p. 175:

Women are in churches, Saints: abroad, Angels: at home, devills: at windowes Syrens: at doores, pyes: and in gardens, Goates.

2) *Down from the waist they are Centaurs,*
Though women all above:
But to the girdle do the gods inherit,
Beneath is all the fiends';
There's hell, etc.
(Lear, IV, vi, 123.)

St. Augustine in his *De Haeresibus,* lxxxv, says: Paterniani inferiores partes humani corporis non a Deo, sed a diabolo factas opinantur, et omnium ex illis partibus flagitiorum licentiam tribuentes impurissime vivunt. Hos etiam Venustianos quidam vocant.—Compare Furness's note, in his Var. Edition.

3) *Isab. . . . . . Go to your bosom;*
Knock there, and ask your heart what it doth know
That's like my brother's fault.
(Meas. f. Meas., II, ii, 136.)

Luther is said to have made an observation to the following effect. If at any time you are not conscious of the sins of the flesh, put your hand into your bosom (greift in euren Busen[1]).

4) The dying Falstaff, the Hostess in Henry V., Act II, iii, 19f., tells us, *cried out 'God, God, God!' three or four times. Now I, to comfort him, bid him a' should not think of God; I hoped there was no need to trouble himself with any such thoughts yet.*

Malone quotes a story from Wits, Fits, and Fancies, etc., 1595, which Shakespeare may have heard: 'A gentlewoman fearing to be drowned, said, 'now Jesu receive our soules! Soft, mistresse, answered the waterman; I trow, 'we are not come to that passe yet'. (Clar. Press ed.)

5) *But earthlier happy is the rose distill'd,*
Than that which withering on the virgin thorn
Grows, lives, and dies in single blessedness.
(Mids. N. Dr., I, i, 76.)

---

[1] Word sused by Bismarck on a memorable occasion.

Compare Sonnet, V, 9—14. In Florio's Second Frutes , 1591, p. 183, we have a passage which throws light on this: "beauty . . . . is also as you "say, a flower sweet, on the stalk, but sweeter in the still."

6) *Here's a farmer, that hanged himself on the expectation of plenty.* (Macbeth, II, III, 5.)

Rabelais (Bk. III, ch. III) says:

'Witness the Usurers of Landerousse, who not long since hanged them-'selves when they saw the Price of Corn and Wine falling and Good Times 'returning'.

Joseph Hall probably had this in mind when he wrote in his Satires, 1598, Bk. IV, sat. VI, 23:

Ech muck-worme will be rich with lawlesse gaine,
Altho he smother up mowes of seven yeares graine,
And hang'd himself when corn grows cheap again.

7) *What is honour? a word. What is in that word honour? what is in that honour? air. A trim reckoning. Who hath it? he that died o' Wednesday*, etc. (1. Henry IV., Act V, I, 135.)

Compare Marlowe's Hero and Leander, I, 269f:

This idol, which you term virginity,
Is neither essence subject to the eye,
No, nor to any one exterior sense, etc.

or Daniel, Complaint of Rosamond, 264f:

Fame (whereof the world seemes to make such choice)
Is but an Eccho, and an idle voice . . . .
. . . Breath of the vulgar . . .
Honor, a thing without us, not our owne, etc.

8) *She is a woman, therefore may be woo'd;*
She is a woman, therefore may be won;
She is Lavinia, therefore must be loved.
(Titus Andr., II, I, 82.)

She's beautiful, and therefore to be woo'd;
She is a woman, therefore to be won.
(1. Henry VI., Act V, III, 78.)

Gentle thou art and therefore to be won,
Beauteous thou art, therefore to be assailed.
(Sonn., XLI, 5.)

---

[1] In this work we also find the pun of 'Stoic' and 'stock' (p. 171) which is used by Lyly and by Shakespeare.

This kind of reasoning is used by Greene:

"Pasylla was a woman, and therefore to be wonne." (Planetomachia, 1585, Grosart's ed., V, 56; comp., too, VII, 68, and VIII, 88.)

9) *Whip me, ye devils,*
From the possession of this heavenly sight!
Blow me about in winds! roast me in sulphur!
Wash me in steep-down gulfs of liquid fire!
(Othello, V, II, 277.)

the delighted spirit
To bathe in fiery floods, or to reside
In thrilling region of thick-ribbed ice;
To be imprison'd in the viewless winds,
And blown with restless violence round about
The pendent world; or to be worse than worst
Of those that lawless and incertain thought
Imagine howling.
(Meas. f. Meas., III, I, 121 f.)

Compare Hamlet, I, IV, 11—22. The reader will, of course, be reminded of Dante, who probably echoes popular beliefs. In Chaucer's *Parlement of Foules,* 78, we read:

But brekers of the lawe, soth to seyne,
And lecherous folk, after that they be dede,
Shul alwey whirle aboute therthe in peyne,
Til many a world be passed, out of drede, etc.

Compare Skeat's note to this in his edition of Chaucer.

10) *Let the brow o'erwhelm it [the eye]*
As fearfully as doth a galled rock
O'erhang and jutty his confounded base,
Swill'd with the wild and wasteful ocean.
(Henry V, Act III, I, 11.)

Compare Nash *Christ's Tears,* 1593, (Grosart's ed., p. 103): 'Like an 'over-hanging Rocke eaten on with the tyde, ... so did theyr prependant 'breastbones imminent-overcanopy theyr bellies'.

11) Glendower. *I can call spirits from the vasty deep.*
Hotspur. *Why, so can I, or so can any man;*
But will they come when you do call for them?
(1. Henry IV., Act III, I, 53.)

Mr. P. A. Daniel[1] illustrates this by quoting from Luigi Groto's *La Calisto* (1580), III, III, where Febo speaking of a great magician, says of him that he can

[1] New Shaksp. Soc. Trans. 1887/92 Part II, p. 246.

'Da gli antichi sepolchri chiamar le anime'—

to which another character replies:

Ben, il chiamarle sarà cosa facile.
Il caso sia, che vogliono rispondere.

The coincidence is no doubt due to intellectual kinship between poets, which must necessarily be the cause of curious parallelisms sometimes.

12) Mr. Reed points out a curious resemblance between Hamlet, II, II, 100f.

Pol. . . . and now remains
That we find out the cause of this effect,
Or rather say, the cause of this defect,
For this effect defective comes by cause —

and St. Augustine, *De Civ. Dei*, XII, 7:

Nemo igitur quaerat efficientem causam malae voluntatis: non enim est efficiens, sed deficiens; quia nec illa effectio est, sed defectio.

This to me is mere accident. Puns were the rage in the Elizabethan age.

III.—Proverbs and Sayings.

I hope a list of proverbs used or alluded to by the great poet will be drawn up in the forthcoming edition of "Shakespeare's Library", the proverbs themselves being given on the left hand side of the page in alphabetical order, while the Shakespearean passages together with a very few other illustrations (wherever necessary) from other authors appear on the right hand side. A tentative list of proverbs has been given by Malone (Var. Ed., XXI) and by Halliwell (Handbook Index). M. C. Wahl wrote 'Programmes' (Erfurt and Leipzig, 1884, '5, '6) on Das parömiologische Sprachgut bei Sh. This work was reprinted in abbreviated form in Shakespeare Jahrbuch, XXII, XXIII.[1] Compare, too, Lippincott's Magazine, No. 58, p. 567, Old Saws of Shakespeare; and Dyer, Folk-lore.[2] Heywood's and Ray's Proverbs should be searched through, also the list in Camden's *Remains*, and in 'Book of Merry Riddles'.

IV.—Folk-Lore and Popular Errors.

T. F. T. Dyer has written a work on Shakespeare's Folk-lore (1884). T. Keightley is the author of 'Fairy Mythology'; and W. C. Hazlitt of 'Shakespeare's Mythology, Fairy Tales', etc. (1875). The following books I have not seen: H. N. Ellacombe, Plant Lore and Garden Craft of Shakespeare 1878; J. E. Harting, Ornithology of Shakespeare (1871). A short paper might be written on popular errors in Shakespeare. Sir Thomas Browne's work against these *idola* should be compared. Shakespeare was, we cannot be

[1] Wahl forgot to state what his abbreviations signify. Who would guess that v=Volksmund? This we ascertain from the Programme.

[2] cf. also Notes and Queries.

surprised at it, the child of his age, and as such held some crude notions. Nor was Bacon in advance of his time. He preached experimentation but he did not practise it. And, we may remark parenthetically, he brought about no revolution in science.

Shakespeare, e. g., believed that a toad was poisonous, and that it contained a precious stone in its head; that a pelican fed her young with her blood; that a snake could sting with its tongue; that toothache was caused by a worm, etc.

V·—Inquiry into Shakespeare's Knowledge of Law, Music, Medicine, etc.

William Blades mentions some works treating of Shakespeare as a musician, as a lawyer, as a chemist, etc., etc., in his 'Shakespeare and Typography', 1872. Lord Campbell and W. L. Rushton wrote books on Shakespeare's legal acquirements. 'Shakespeare and Music' is the work of Mr. Naylor; and Chappell's 'Popular Music' is the best book on early English music. Many articles have been written on Shakespeare as a physician. But what has he to do with medicine? An interesting paper (not a large book) might be written, in which some sayings of Shakespeare were viewed through the spectacles of the old physicians. Galen, Paracelsus, Hippocrates (all mentioned by Shakespeare) were the medical authorities then.[1]

[1] Perhaps T. Lund, Gesundheit und Krankheit in der Anschauung alter Zeiten, 1901, might be found useful.

I need hardly say, in conclusion, that the absence on the preceding pages of names such as W. Hazlitt, Gervinus, Moulton, White, Brandes, Engel, Franz, Kreyssig, Wetz, etc., etc., familiar to the students of Shakespeare, is simply due to the fact that their sphere of work lies in another direction.

INDEX.[1]

[1] The abbreviation Sh. means Shakespeare; *n* = note, reg. = regarding; s. v. = sub verbo; etc. I have not indexed the geographical names on pp. 250—256 (cf. p. 271) and 237 *n* 2 and the plays where they occur. *A propos* of Wincot (p. 254), cf. S. Lee, Life of Shakespeare, p. 165.

Bei Fragen zur Produktsicherheit wenden Sie sich bitte an:
If you have any questions regarding product safety,
please contact:

Walter de Gruyter GmbH
Genthiner Straße 13
10785 Berlin
productsafety@degruyterbrill.com